Data for Agrarian Development is a comprehe
and managing agricultural data in developing
general application to people working in the p
management of tropical agriculture. Many dev
the need to collect data for planning, monitori
project, or to use data collected by other people. This book explains how to plan, design and manage a survey, and introduces basic analysis and report writing. Unlike other books in this field the authors deal with specific statistical issues of simple and multi-stage sample design and analysis. In this way the book bridges the gap between statistics texts, which are too mathematical for many readers, and general development books which omit statistical details. An understanding of basic techniques of sampling and data collection will enable those people who have to collect data to improve its quality and permit careful appraisal of data from other sources. The book is intended for non-mathematical readers; simple practical examples are used throughout to illustrate the methods recommended by the authors.

Data for agrarian development

WYE STUDIES IN AGRICULTURAL AND RURAL DEVELOPMENT

Solving the problems of agricultural and rural development in poorer countries requires, among other things, sufficient numbers of well-trained and skilled professionals. To help meet the need for topical and effective teaching materials in this area, the books in the series are designed for use by teachers, students and practitioners of the planning and management of agricultural and rural development. The series is being developed in association with the innovative postgraduate programme in Agricultural Development for external students of the University of London.

The series concentrates on the principles, techniques and applications of policy analysis, planning and implementation of agricultural and rural development. Texts review and synthesise existing knowledge and highlight current issues, combining academic rigour and topicality with a concern for practical applications. Most importantly, the series provides simultaneously a systematic basis for teaching and study, a means of updating the knowledge of workers in the field, and a source of ideas for those involved in planning development.

Also in the series

Frank Ellis *Peasant economics: farm households and agrarian development*
Niels Röling *Extension science: information systems in agricultural development*
Robert Chambers *Managing canal irrigation: practical analysis from South Asia*
David Colman and Trevor Young *Principles of agricultural economics: markets and prices in less developed countries*
Leslie Small and Ian Carruthers *Farmer-financed irrigation: the economics of reform*
Frank Ellis *Agricultural policies in developing countries*

Data for Agrarian Development

C. D. POATE & P. F. DAPLYN

Published by the Press Syndicate of the University of Cambridge
The Pitt Building, Trumpington Street, Cambridge CB2 1RP
40 West 20th Street, New York, NY 10011–4211, USA
10 Stamford Road, Oakleigh, Melbourne 3166, Australia

First published 1993

Printed in Great Britain at the University Press, Cambridge

A catalogue record for this book is available from the British Library

Library of Congress cataloguing in publication data

Poate, C. D., 1950–
Data for agrarian development / C. D. Poate & P. F. Daplyn.
p. cm. – (Wye studies in agricultural and rural development)
Includes bibliographical references and index.
ISBN 0-521-36566-X. – ISBN 0-521-36758–1
1. Agriculture – Economic aspects – Statistical methods.
2. Agriculture – Statistical methods. 3. Rural development –
Statistical methods. 4. Agriculture – Economic aspects – Data processing.
5. Agriculture – Data processing. 6. Rural development –
Data processing. I. Daplyn, P. F. II. Title. III. Series.
HD1425.P6 1992
338.1′072 – dc20 92-22831 CIP

ISBN 0 521 36566 X hardback
ISBN 0 521 36758 1 paperback

PN

Contents

Preface

For many years we have both been interested in the use of statistical data by people working in agricultural development. The need for data for planning and evaluation has encouraged development workers to carry out surveys as a routine exercise. Yet even the smallest survey requires care in its planning and statistical design. In our own exploration of the text books available to help beginners we were struck by an apparent polarisation: there were books with a broad treatment but virtually no mathematical coverage of sample designs and their formulae for analysis; and other books with rigorous maths, but little in the way of practical advice. There was clearly a need for a middle course, something with general descriptions for the non-mathematical reader, but with specific examples of the maths for those who are interested.

The genesis of this book was in work done while we were both working for the Agricultural Projects' Monitoring, Evaluation and Planning Unit of the Federal Government of Nigeria, and found a need to produce a compendium of information, both theoretical and practical, on surveys of farms and farmers. We are indebted to many colleagues, project field staff, farmers and survey respondents who contributed ideas, information and insights to this work.

The scope and coverage of this book are discussed in the Introduction. We hope that it may prove useful to a wide audience of workers in agricultural development who need to collect and use data in the course of their work. It was, however, written primarily for students of the External Programme MSc degree at Wye College. We have accordingly received in the course of writing much helpful comment from staff of the External Programme, and are particularly grateful to Drs Henry Bernstein, Cecile Jackson, and Hassan Hakimian.

The writing and editing of the manuscript has taken much time away from our work colleagues and families. We owe them our thanks for their patience and encouragement.

We are indebted to Dr Michael Daplyn for the original ideas and examples of Appendix 2.

And finally, we would like to express to each other our appreciation for the judicious combination of encouragement and harassment which, mutually applied, has resulted in the completion of this book.

Introduction

Scope and coverage

The reader

This book has been written for a number of different readers. Primarily it is intended for the students of the External Programme degree at Wye College. The book contains much of the material for the core course on Economic and Social Survey Methods and Data Analysis. It assumes that students have little or no practical experience in data collection but have completed a basic course in statistics and understand the concepts of central tendency and dispersion. Readers should thus be reasonably familiar with the calculation of mean and variance from a set of data and with the algebraic notation used for basic statistical formulae, and have a general understanding of the concepts of probability, and hypothesis testing.

Although some statistics are necessary to follow the material, the mathematical aspects of surveys takes up a relatively small part of the book because our intention is to present the survey process as a whole. Surveys are practical exercises and we have tried to concentrate on techniques and methods of working that can be applied by development workers who are not trained statisticians, but whose work requires them to collect data of some kind. For this reason we hope the book will also appeal to a wider audience of practitioners in the field of agricultural development, who occasionally collect data and often make use of other people's data.

The growth of monitoring and evaluation activities has given a new impetus to rural surveys in recent years. Countless farm and village surveys take place every year in developing countries. Some are designed as bases for project planning; others are for academic or management-oriented research. But many, if not most, of the surveys are poorly designed, badly collected, never documented, and reported months or years later than expected. It is our belief that these problems often arise from a misplaced conviction that data collection is not only an easy task but also one which any professional in agriculture, economics, sociology or similar development-related subjects, can embark on without further training. It is not. Good survey management calls for a team with an exceptional mix of talents: high standards of professional competence to define survey objectives and information needs; a sound grasp of statistics to ensure that the data represent the population from which they are drawn and can be analysed simply; a flair for detail and accuracy when handling numbers; administrative skills to draw up and control survey resources and work programmes; the ability to motivate field staff and create a rapport with respondents and officials; the energy to adhere to deadlines and keep the process moving; and last, but not least, the interpretive and analytical skills to prepare a literate, concise, report.

We do not claim that this book will turn the reader into a survey specialist, but it is hoped that it will offer sound, effective techniques which can be used to achieve high standards of data quality. For this reason we hope the book will appeal to development workers at two different levels: the novice surveyor looking for techniques and guidance for a new survey; and the reader or commissioner of a survey who wants to know what to look for in standards of documentation, survey design, methodology and data processing, before accepting, or believing, a survey report.

The coverage

Our aim is to present the survey process as a whole – by which we mean both the sequence of activities from planning to reporting, and the breadth of alternatives in design, in scale of operations, and in methods. Of necessity, coverage is restricted to the most common statistical designs and data collection methods. For these, we have tried to present sufficient detail and examples for them to be employed without further reference. However, it must be stated at the outset that although the book takes a practical stance, it is not a manual and leaves the reader to make his or her own decision about the best approach for a particular survey.

As part of the concept of choice we wish to establish two principles which are frequently misunderstood. First, that there is no single correct survey method. In most instances there is a variety of alternatives involving the survey statistical design, the scale of operations and the methods by which data are collected. A good design will match the characteristics of different approaches to the objectives and constraints of the survey. Some examples of this are given below. Second, that statistical design, scale and technique are to a large extent independent of each other. Many preconceived ideas exist here: that sample surveys always involve large numbers of observations; that case studies are not based on random sampling; that crop cutting is the best way to measure crop yields; that rapid appraisal does not involve structured questioning. None of these statements are necessarily true. A sample survey can have a handful of observations; a case study can be carried out on a group of respondents chosen at random from a defined population; more useful yield figures can sometimes come from farmers' own estimates rather than crop cutting; and rapid investigations can make use of a combination of structured and unstructured interview techniques. The surveyor must match the approach to the purpose.

Although we argue that different approaches are suitable according to particular circumstances, the main thrust of the material in the book is based on formal sample surveys. There are three reasons for this. First, we believe that formal techniques such as random sampling and structured questionnaires have a wider range of applicability, and if applied accurately, they are capable of producing more reliable information. Second, there is a greater body of theory and established procedures for formal techniques, which when understood, permit knowledgeable adaptation for informal studies. Third, and related to this, the two procedures are by no means mutually exclusive: less formal appraisals can often be used as a means of exploring the grounds for more formal surveys. The readers should, therefore, feel free to consult, at the same time as reading this book, a growing body of literature on informal surveys.[1] We firmly agree with the need to choose the most cost– and time-effective method for the problem at hand. What matters most is to know when and to decide how to apply each procedure.

Among managers and administrators, financial and physical information are regarded as 'hard' data (accurate measurements); most survey findings are considered to be 'soft' (interpretations or estimates). This distinction also runs through our discussion in Chapter 4. Financial records are facts, not estimates. The information reported from a survey, on the other hand, will vary from a simple physical measurement (a hard fact if accurate) to an enumerator's interpretation of a respondent's answer to a question. If the question was wrongly phrased, the answer misunderstood, or even the wrong respondent questioned, the information reported will be different from reality (therefore very soft or unreliable). The more the subject matter requires close and skilled questioning, and careful interpretation, the harder it is to get accurate results. Most agrarian surveys contain a range of topics, from the hard (measure the area of a field of cassava) to the soft (in your opinion, why were yields lower than last season?).

In this book we concentrate on the 'hard' topics for data collection. Most examples are drawn from survey topics which involve measurement, or measurement plus interviews. The book does not deal with qualitative data collection or opinion surveys. Neither do we cover applied procedures, such as diagnostic or reconnaissance surveys for farming systems research. Some of the techniques employed by such surveys are the same as the techniques for formal surveys – selection of units of study, drawing up questions, designing a data collection form, data processing and analysis. It is the context in which the survey is used which determines the procedure to be followed. This important point must be borne in mind in the planning and management of any survey programme.

Structure of contents

The book is divided into three parts: preparation; the survey; and understanding the data. These are intended to encompass most aspects of survey planning, management and analysis. This can be shown with the aid of a flow diagram, a medium used extensively throughout the book to illustrate relationships and interactions between different topics.

Survey design is a compromise between objectives and resources. This fundamental compromise leads to a choice of method.

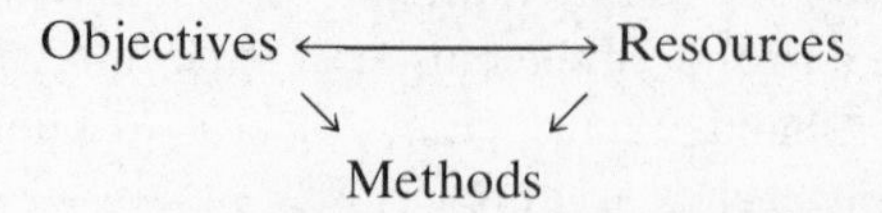

Objectives include the information inferences sought from the sample to be surveyed and topics to be covered. To start and plan a survey, however, we also need clear objectives in terms of accuracy (how close to the truth must the information be), precision (within what confidence interval are the findings requested), and timing (how soon are the results needed). This in turn raises the important question of the resources a surveyor has at his or her disposal. Available finance, level of enumeration skills and experience will determine the scope of work which can be undertaken. But timing is also critical: with adequate resources a survey team can be trained from scratch, otherwise the existing set-up must do. The method or approach will follow from a review of objectives and resources. Later, in Chapter 1, we summarise rapid techniques, case studies, sample surveys, census and remote sensing. In most studies no single technique will be enough. To illustrate the need to combine methods in practical work, section 1.3 describes three surveys and discusses considerations leading to choice of method.

Preparation

Once a broad approach is decided upon, survey planning must begin. The aim here is to consider detailed aspects: survey method; sample size, selection and location; data collection techniques; enumeration requirements and existing skills; and seasonal or other timing considerations. These factors combine to determine the accuracy, precison and timing of the results. If these do not meet the survey objectives then either the objectives must be reviewed, or the planning process repeated in an iterative manner until a satisfactory approach is determined. Planning is considered in Chapter 1, with emphasis not just on the survey design, but also the timing of envisaged activities and aspects of data processing. Sampling concepts and design are considered in Chapters 2–3, where we include details of some of the most widely applicable and straightforward schemes of selection. Sample selection may be purposive with no attempt at randomness. The argument as to why this is

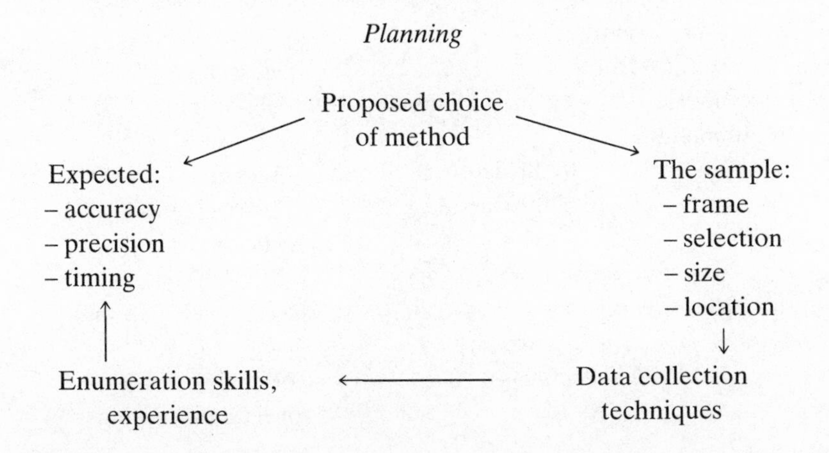

rarely the best approach is an important one and is explored in Chapter 1. Data collection methods are reviewed in Chapter 5, with emphasis on factual data items.

The survey

To many readers the survey consists of the questionnaire and methods of data collection. We hope the material here will broaden that view to include management and analysis, even though data collection remains a central topic. Collection and questionnaire design are closely interlinked. Collection is concerned with principles of interview and measurement and the specific techniques which can be applied in gathering information. Chapters 4–7deal with various aspects of data collection. Techniques of collection are largely independent of the medium on which the responses are recorded. Chapter 7 looks at questionnaire design, with emphasis on the non-verbatim style of layout which is most suitable for trained enumerators.

Conducting a good survey, however, is contingent upon a demanding mixture of professional detail and practical management. In this book, we try to give due prominence to management because so many well-designed surveys fail when their implementors forget to follow painstaking design with energetic field visits and support for their enumerators.

The survey

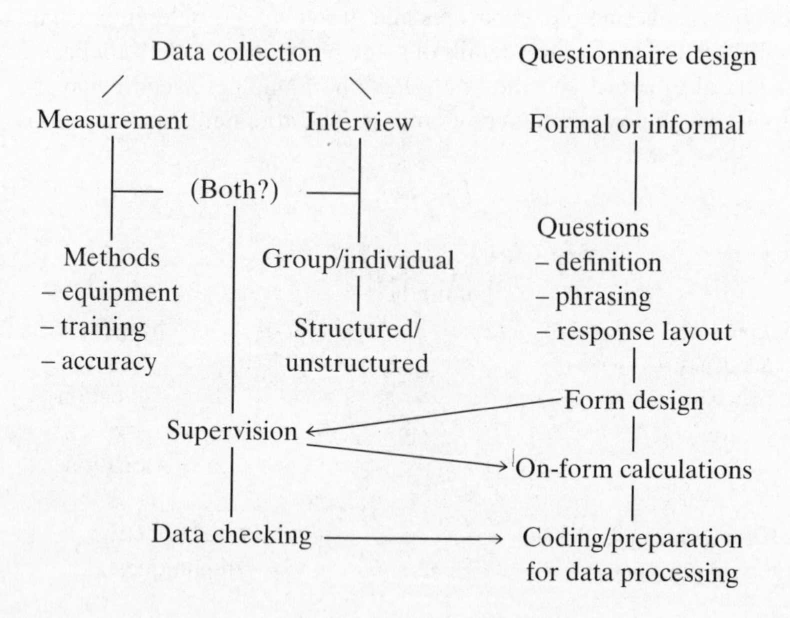

Fieldwork management is an essential part of such quests for good data quality. It is both internal, concerned with the timetable of work, training and documentation; and external, dealing with government agencies, other surveyors, community leaders, and the welfare and performance of the field workers. Specific issues about field recruitment and training are set out in Chapter 8. We have tried to be comprehensive in our coverage of enumerator selection and training, documentation, publicity and field operations. Although the material is intended as a set of guidelines, inevitably some of it may appear prescriptive, even implying standards that in practice are hard to achieve. This is acknowledged, but we hope the reader will be selective in applying ideas to those aspects of his or her own survey where they will be most effective.

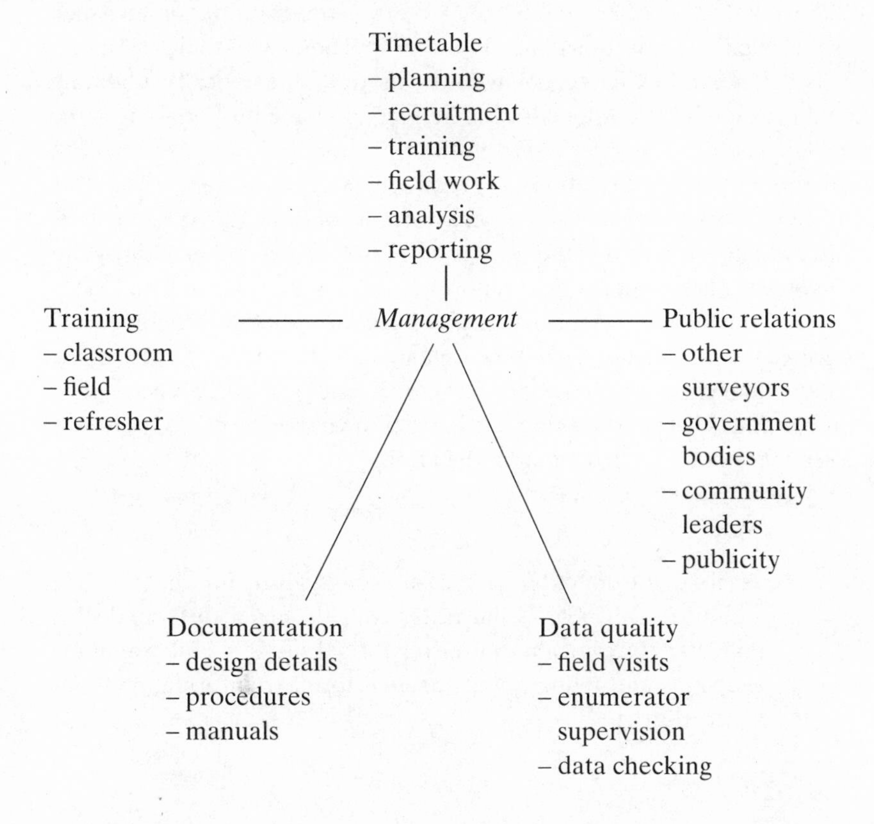

Analysis

Once the data have been collected they must be transformed into usable information. We identify three stages, dealt with in Chapters 9, 10, 11 and 12. These start with data processing: the hand or computer

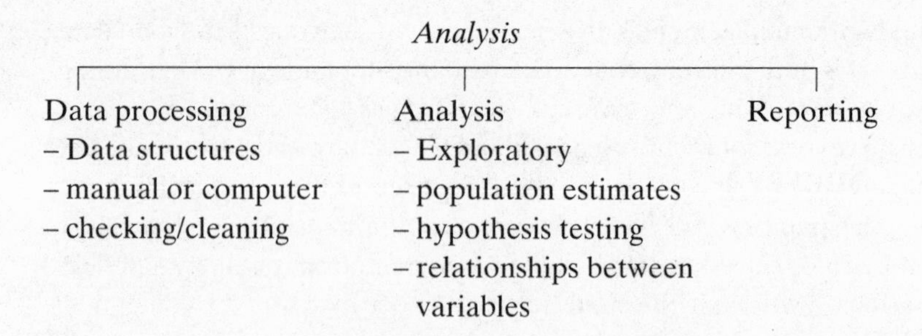

operations to list, validate and summarise the raw data, in order to check its content and completeness. There then follows analysis proper at several different levels. First, an exploratory analysis, to identify the characteristics of variables and to make certain population estimates such as the calculation of totals, means and proportions with their variances. This is covered in Chapter 10, with the more mathematically inhibiting detail gathered into Appendix 1. The second element of analysis is the techniques for testing certain hypotheses regarding differences between subpopulations, and relationships between variables (Chapter 11). This section makes use of techniques and concepts of basic statistical analysis with which we assume the reader is reasonably familiar. Lastly, in Chapter 12, are guidelines for report writing and presenting data.

While the overall structure of the book conforms to the chronologically sensible layout as described above – planning, management, and analysis – individual chapters are nevertheless fairly independent. Some readers may wish to select certain chapters for their own specific needs, or follow a sequence different from that in the book.

Notes

1. A good introduction to the subject is given in Chapter 3 of Chambers, (1983). It illustrates well the need for sound data collection by critically appraising formal surveys, and examining the skills and requirements needed to make informal methods successful.

1

Survey planning: objectives, methods and constraints

1.1 Survey planning

Planning is a compromise between objectives and resources, an iterative process in which an ideal outcome is modified to suit the practicability and costs of actual data collection. This book takes the reader through the stages of survey preparation, execution and analysis, in sequence. However, like many iterative processes, decisions about data items, accuracy, means of collection, sample size, and timing are inter-related, change one and the change affects the others. So where do we begin? A good tale begins in the beginning, but a survey begins at the end.

Design

Setting *objectives* involves a dialogue between the 'client' for whom the survey is being conducted, and the surveyor. At the very least this must include a specification of *topics* and items of data. But this alone is likely to be insufficient. The best approach is first to prepare a draft statement summarising the results needed from the survey. From these can be set out dummy tables for the final report, which show the relationships of interest between variables. If necessary, this should include tests of hypotheses. This is often a difficult exercise, but one which repays the effort handsomely, for it leads to accurate specification of the survey objectives.

A simple example can illustrate this. One of our first concerns will be the *unit of study* for data collection. (For a definition of technical terms, see the Glossary at the end.) A survey of crop production which aims to compare performance between farmers in different localities, or contrast farmers with different social characteristics, can be based on the farming household. But if the objective is to test the results of a new seed variety, or of fertiliser inputs, information will need to be based on farmers' fields where separate treatments can be distinguished. A family may employ different strategies on different fields, so household-level data would not portray the treatment effect.

This distinction may enable us to redefine the *universe* or *population* for study. Often populations are accepted in natural groupings: the farming community in a region, all the rice grown on an irrigation scheme. In fact, the real interest may only be farmers who take credit, or fields on which fertiliser was applied. This awareness helps us to make the survey more specific, and therefore more likely to meet its objectives. Surveys carried out to monitor progress under a project may not require that population estimates are produced from sample data. It may be sufficient to compare samples from specific groups with each other: adoption of recommendations by contact farmers compared with non-contact farmers. By specifying the universe in this way, the compilation of a *sampling frame* can be simplified and it may be possible to use a simpler statistical design.

If it is necessary to test for hypotheses, especially concerning differences between sub-populations, or in comparison with population norms, the issue of *precision* will arise. To know that an adoption rate is 40% may be valuable, but what precision does the sample size bring? 40% with an accuracy of plus or minus 15% is less useful than 40% with an accuracy of plus or minus 2%. The interest may not be in estimating change from one round of the study, but to record the trend over a period of years. The ability to maintain data collection over an extended period, and to achieve the levels of precision necessary to demonstrate a convincing trend, are better judged before the survey is designed, than after several years of collection. With an understanding of the precision needed, the desired sample size can be estimated.

Measuring any change usually implies a comparison between sub-groups of the population. Trying to infer causality, such as extension advice leading to higher crop yields, is beset with difficulties when experimental techniques cannot be utilised. Control groups may be an option, and these would need the same specification and selection as for the population of interest.

Knowing precisely who to survey is the first stage in planning. What to collect comes next. A wide range of *methods* exist from rapid assessment to repeated visit collection. Within this range the appropriate level of *measurement* or *interview* needs to be chosen. A study to quantify the impact on farm labour utilisation of a new technology may necessitate detailed recording of labour patterns and work rate. Alternatively, a quick assessment of farmer adoption of that technology may find their perceptions of changing labour patterns sufficient. What combination of methods should be used? It is unlikely that any single method would be suitable for all the data required.

The choice of design (and scale of enquiry) will also depend on *enumeration resources*. Skilled enumerators might undertake less structured, open-ended interviews, guided perhaps by checklists of topics. Inexperienced field workers, or large teams for large samples will tend to need more structured instruments, carefully phrased to ensure consistency of enquiry, and perhaps more limited in scope.

Data processing

It is important, yet often neglected, to integrate *data processing* into the planning process from an early stage. The content and methodology of the survey affect the data processing requirements and are in turn affected by them, especially concerning the sample design, data recording and structure. A major choice to be made at an early stage is whether to use manual or computer processing. The widespread availability of microcomputers, at relatively low cost, means that computers are increasingly likely to be the preferred option. However, for small enquiries manual processing can still be attractive, especially if one objective is for the enumerators, or even respondents, to participate in the process. With either method, statistical clerks or keyboard operators may need to be recruited and trained, and processing procedures designed and documented.

The decision to use a computer needs to be taken early, because it affects and constrains many other aspects of the survey planning, including the content and layout of the questionnaire, the sample design and size, the checking and editing procedures, the analytical procedures and methods, and the reporting schedule. It is thus important that this decision be made in the context of the planning of the survey as a whole.

Where the processing and analysis are to be done by computer, it is important to decide upon the necessary hardware, software, and personnel, and ensure that they will be available at the appropriate time. The issues involved in the selection of hardware and software are discussed in Chapter 9.

Planning Checklist

A planning checklist will help ensure that certain items and activities are not overlooked. As a starting point for your planning, you may draw up a checklist along the lines of the example below. Generally, you should try to complete it before moving on to the next stage.

EXAMPLE 1.1
Planning checklist

1. **Statement of objectives**
2. **Topics to be covered**
3. **Content of tables needed in the final report**
4. **Unit of study (is more than one necessary?)**
5. **Population to be studied**
6. **Sample frame (available, or to be constructed)**
7. **Necessary precision – sample size**
8. **Ability to measure changes over time**
9. **Need to infer causality**
10. **Methods to be used**
11. **Enumeration resources and experience**
12. **Questionnaire style**
13. **Data processing system**

Timetable

Once there is an agreement to proceed with the survey, a planning timetable should be drawn up in order to facilitate planning and budgeting. Start by working backwards from the date when data collection will start. (As mentioned above, the planning of data processing will also need to be considered at an early stage, but the discussion here will concentrate on pre-fieldwork timetabling of activities.) Be realistic about the time necessary to complete each stage of the work, especially where approval must be sought from other organisations, or colleagues have to be consulted.

The activities should be listed against their target dates as in Example 1.2. If you find it useful, a bar chart can be constructed to represent the duration of each task. This can be particularly worth doing when used to monitor progress by marking actual work done against the plan.

Example 1.2 illustrates a lengthy period of preparation for a large-scale survey which will last over one whole season. Surveys of smaller scale or

EXAMPLE 1.2
Agronomic survey planning timetable

For a survey starting in March designed for crops with a growth season of March/April to November/December.

April	**Define broad objectives**
	Identify senior staff for survey
	Form Advisory Committee
May–July	**Draw up outline plan for:**
	objectives
	universe
	precision
	types of measurement
	sample size and design
	questionnaire
	field staff
	resources and equipment
	funding
	data processing
	analysis and reporting
	Draw up outline budget
August	**Draft forms and manual**
September	**Draw up sampling scheme**
October	**Draw up detailed budget**
	Draw up processing and analysis scheme
November	**Place printing orders for forms and manuals**
	Place equipment orders
	Start recruitment arrangements for extra field and office staff
	Make first-stage sample frame
December	**Plan training programme**
	Test and interview enumerator candidates
	Plan publicity campaign
	Start arrangements for field staff accommodation
	Start arrangements for transport
	Do first-stage sampling
January	**Procure stationery**
	field equipment
	data processing equipment
	forms and manuals

continued

January *continued*	**Make advance publicity arrangements**
	New staff start work
	Make second-stage sampling frames
February	**Train field staff**
	Do publicity campaign
	Do second-stage sampling
	Distribute materials and equipment
March	**Start field work**

shorter duration may require less preparation, but not proportionately less. Design of forms and manuals, recruitment and training of staff, and logistics arrangements are all time-consuming and may take as long for a survey lasting one month as for a survey lasting six months.

Costs

Whilst at an early stage in planning, *costs* need to be assessed against the available budget, or a budget proposal needs to be prepared. The first step is to compile a checklist, such as that illustrated in Example 1.3.

Surveys, with a high proportion of investment and fixed costs, are expensive. Survey equipment, such as vehicles, measurement instruments, and computers have a working life beyond a typical single study. The creation of a one-off survey unit may well be financially unattractive if more permanent options are available. Professional and field staff are usually hired on a permanent, or long-term basis, to bring continuity and security to their working arrangements, and thus invest in their capabilities. Their salaries will be fixed costs. Variable costs tend to be restricted to discrete items, such as training courses, or costs directly associated with contacting respondents: vehicle operating expenses, accommodation, night allowance, and daily paid survey assistants. For this reason, there is often a temptation to expand the scale of data collection to the limit of enumeration resources, to spread overhead costs and maximise statistical precision, but large scale can bring problems with data quality.

During the survey it will be necessary to monitor expenditures – a simple analysis against budget items on a spreadsheet will do – and for a system to facilitate cash expenditure on minor consumable items such as stationery or repairs to equipment.

EXAMPLE 1.3
Agronomic survey: items to be budgeted for

Personnel	**enumerators, salaries and allowances**
	supervisors, salaries and allowances
	field officer, field allowance
	survey officer, travel expenses
	statistical clerks, salaries
Training	**hire of dormitories, classroom**
	food
	transport to field sites
	hire of video-recorder and television
Transport	**landrovers, running costs**
	motor cycle allowances
	bicycle allowance
Accommodation	**enumerators, housing**
	supervisors, housing
	office, store
Equipment	**prismatic compasses**
	measuring tapes
	weighing slings
	weighing scales
	moisture meters
	sacks for crop produce
	clipboards
	calculators
	microcomputer, hardware, software, consumables
Stationery	**questionnaires, plus spares**
	manuals
	reporting pro-formas
	document control forms
	stickers
	pens, pencils, sharpeners, erasers, rulers
	computer listing paper
	report production
Publicity	**posters**
	leaflets
	loudspeaker vans
	hire of rooms for meetings
	radio announcements

Planning compromise

The skill in planning is to match objectives with resources through adjusting methods used to achieve them. Often there exists a wide range of choice in the techniques used for data collection: complex measurements versus simple observations, versus open-ended interviews; the possibility of a single visit to the unit of study, versus repeated rounds of enquiry. Enumerators' costs are influenced by their skills and training, and also by logistics. Some studies require enumerators to be resident at the site of data collection. A mobile system might permit the enumeration of the same sample size using fewer people, but costs will depend on the mode and frequency of transport.

Undoubtedly, the main feature for compromise is often the sample size. There may be scope to work with a smaller or less scattered (clustered) sample by changing the population to be studied, such as fewer geographical regions. But ultimately the absolute size of the sample is likely to have the biggest impact on costs. Sample size is discussed in Chapter 3, where technical issues relating to sampling are considered. For general guidance however, we believe that a smaller sample size yielding good quality data is preferable to the alternative of a larger sample size retained in pursuit of greater statistical precision. As we shall see, resource constraints can have a limiting impact on the enumeration workforce, training period or duration of the survey, hence increasing the possibility of non-sampling errors.

1.2 Choice of design

Types of enquiry

Several types of enquiry for the gathering of data on agricultural development topics can be distinguished, some of these are described below:

(a) *Census*. This is an investigation which covers every individual unit in the population being studied. The best known examples are the national censuses of population and of agriculture which are conducted by many countries on a regular basis. A census can, however, relate to a much smaller and more specific population, for example, a population census of an individual village; or a study of all farmers receiving credit from an agricultural bank; or an investigation covering all extension workers in a district. These three examples reflect the two general circumstances in which a census is likely to be appropriate. One is where basic information is needed for general planning purposes, so that it is necessary to gather data on a range of topics, from all sectors of the population with a good level of accuracy. The other is where the population under investigation is small and readily identifiable, but may be fairly heterogeneous, so that it is both feasible and advisable to include the whole of the population in the study.

(b) *Sample survey*. This is an investigation in which only part of the population is studied. Usually, but not always, the selection is made at random, so that the sample is known to be representative of the whole population. This type of enquiry is appropriate when resources are not sufficient to cover the whole population, and the lower level of precision involved in covering only a sample is acceptable, but the information gathered must be able to be generalised to the whole population. A sample survey may be for general planning purposes, such as a national demographic or labour force survey, in which case it is a rather large-scale operation. This is not necessarily always the case: a sample survey can be of any size. For example, the take-up of an extension package might be investigated using a sample survey of farmers in a project area, or the effect of the rehabilitation of a watercourse might be assessed using a sample survey of farm holdings lying along that watercourse.

(c) *Aerial survey*. An aerial survey may, formally, be a census or a sample survey, but (in either case) it has such particular characteristics that it needs to be mentioned separately. This type of survey involves an aircraft flying over the whole or a part of an area of interest, gathering information about the ground below either by taking photographs or film, or by an observer recording his or her observations while in the air. (Remote survey by satellite also falls into this category.) It is therefore suitable for gathering information only on topics which can be investigated by direct observation, without interviewing or otherwise contacting a respondent, and which are distinctive enough to be seen from the air. Some examples of topics which may be appropriate are numbers of cattle or large wildlife, land cultivation categories, cropping patterns where the different crops are very distinctive, settlement locations and approximate sizes. Aerial surveys are particularly useful for obtaining, very rapidly, broad information about large areas.[1] They are, however, expensive, in terms not only of money but of skilled manpower and sophisticated equipment.

(d) *Case study*. This is an enquiry in which a small number of study units are investigated in great detail. The units selected are not necessarily on a random basis, but by virtue of having certain characteristics which are needed for the investigation. The focus of a case study is on the detailed structures, patterns or inter-relationships observed within each individual case included in the study, though the cases themselves may be selected to cover a range of different types of study unit. An example of a case study is often found in farming systems research, at the stage of on-farm testing of research station findings. Another example might be an investigation of the allocation of rural women's time to different activities, using a detailed time budget approach, in order to arrive at appropriate definitions and classifications of economic activity. Here a small number of women, selected to cover a range of types of household and of farm and non-farm activity patterns, would be appropriate.

(e) *Rapid methods*. This category encompasses a variety of investigation techniques used to obtain rapid, not very precise, and sometimes non-quantitative information. They are used in situations either when little time and resources are available, and limited information is still useful, or when no data are initially available, and a quick preliminary enquiry is necessary to provide direction for further study. These methods typically involve an investigator or team working in the study area, observing the characteristics of interest. The observation may be direct or indirect, for example, quantification of the crop mixtures found in the area, or of the incidence of crop failure due to pests or drought. It may alternatively be by interviewing informants, either selected at random or chosen as being knowledgeable about the subject of study – local leaders, prominent farmers, or the elderly of a community, are possible examples. Instances of this method are found, applied for different purposes, under a variety of names, including beneficiary assessment, rapid rural appraisal and reconnaissance surveys.[2]

(f) *Experimentation*. This is a type of enquiry in which a stimulus is applied to a subject, and the effect observed. In classic agricultural experimentation, the stimulus, or treatment, is applied to one group of subjects. A second, control group, does not receive the treatment, so that the effect of the treatment can be assessed by comparing the two groups. All other factors affecting the two groups are held constant so that differences between the control and treatment groups can be related to the treatment being applied. For this approach to be valid, the subjects must be assigned to the treatment or the control group at random. A considerable body of theory has been developed on the design and analysis of experiments. But for the development situations considered in this book, it is not generally possible for the stimuli being investigated to be assigned at random, and other methods must be used. The subject of experimentation is therefore not further considered here.[3]

Design, scale and method

When planning a study it is important to distinguish between the design of an enquiry, the scale on which it is conducted, and the methods used to gather the data. Several types of enquiry are described in the previous section, ranging from the full census to the case study to the rapid appraisal. Generally speaking, these designs are different in kind, not in degree – they are essentially different approaches to obtaining information.

Some of these designs are of course more suitable than others for investigations of different scales or magnitudes. For example, a case study approach requires a very heavy resource commitment for each case: so it is only suitable for use on a small scale. But there is no general or inevitable relationship between the design and the scale of an enquiry. A census, for example, may be a very large-scale enterprise covering every person in a country, or it may be a very small-scale investigation covering every person in a newly formed credit scheme, or every farmer cultivating land along a recently rehabilitated irrigation watercourse. A sample survey may be for the purpose of obtaining a general base of information on farming practices for a whole country, and may cover 100 000 farm households, or it may be for the purpose of obtaining an average price for maize meal in a particular market, and may cover five traders. Scale is an independent issue from design of an enquiry, though some combinations are more likely than others.

A third independent issue is that of the method or methods of data collection used in an enquiry. These may involve direct measurement or observation, or the extraction of information from existing records, or interviewing respondents either using a formal questionnaire, or using a checklist to structure the interview partly, or using a completely open and unstructured approach. These methods are not mutually exclusive, and combinations of methods are often found, perhaps using different methods for different topics, or at different stages of an investigation. It is this flexibility of combination that brings greater scope for adjustment during planning. Some examples are given in section 1.3.

Equally, particular methods of data collection are not necessarily associated with particular designs or scales of enquiry. A large-scale sample survey is likely to involve the use of a formal questionnaire, but so might a small-scale case study. A rapid reconnaissance survey might involve completely unstructured interviews, or it might involve a very detailed and structured checklist.

An enquiry of the rapid appraisal type is sometimes known as an informal survey. This is taken by some investigators to imply that every aspect of such an enquiry is as informal as possible; that all sampling involved must be on an informal or purposive basis, and random samples must never be used; that interviews must be open ended and questionnaires never used. But this does not necessarily follow, and no useful purpose is served by restricting one's options by the terms of any rigid blueprint. The objectives and constraints upon the investigation determine the design, scale and method to be used, in whatever combination best suits the survey purpose.

The need for flexibility in choosing and fitting together the different aspects of an investigation can be seen in the case where a sequence of enquiries is used. This is common where little information is initially available, so that some preliminary, perhaps approximate, data are needed as quickly as possible. An exploratory study is therefore needed first of all to give a general idea of the situation and to identify topics or hypotheses needing further investigation. At this stage a rapid appraisal approach, or perhaps an aerial survey, is likely to be the most suitable.

The stage which typically follows on from this is to consolidate and generalise the information obtained in the exploratory stage. Depending upon the individual circumstances, a census, or sample survey, or some combination of these, may be appropriate. The scale and methods to be used must of course depend on the nature of the information needed, and the resource constraints, as noted above.

A third stage of investigation which may then arise is to examine particular topics in greater depth. Here a sample survey or case study approach is most likely to be suitable, and probably on a small scale. An integrated sequence of studies, operating at different levels of detail and precision, may thus be the best approach. As the objectives vary over the different stages of the investigation, so too do the methods and approaches used to meet them.

1.3 Examples of study designs

So far we have concentrated on a number of important ideas: that in most real surveys a mixture of approaches will be necessary; that sample design, scale of enquiry and choice of data collection methods are largely independent of each other; and that the choice of approach is a compromise between objectives and resources. In this section we illustrate these ideas by means of three examples of the sort of data collection activities this book is designed for. Some of the techniques have already been touched on earlier in this chapter. Others are the subject of later chapters. In case of difficulty with some of the terminology used, the reader should refer to the glossary at the end of the book for quick clarification. Some readers may still find that the examples deal with too many unfamiliar issues at this stage – if so, we suggest they skip them now and return later. A more in-depth treatment of techniques will obviously have to await the coming chapters.

1. *A planning study*

A planning department of a ministry of agriculture and natural resources in a developing country is given the task of preparing a feasibility study for a new irrigation scheme. This will be part of a comprehensive project, which includes studies into ground and surface water, soils and geomorphology, current and proposed agriculture, and engineering of a proposed dam and system of canals. The planning department decides that a formal survey programme is required to assess the social and agro-economic characteristics and potential of the study area. A set of information objectives are set up for the study.

- Location and description of all settlements within a 25 km radius of the study site.
- Total human population and its distribution within the same area
- An assessment of current land use in the area
- Details of households in the area (all are believed to be farming) to cover:
  - household composition
  - farming activities
  - off-farm activities
  - farm output

crop and livestock disposal
details of existing crop systems on seasonally inundated land with a high water table

- Wholesale and retail prices for local commodities and inputs at markets within and adjacent to the study area and official and black-market prices for government-controlled farm inputs

As a benchmark or baseline estimate, these data would permit farm and household budgets to be prepared to calculate financial returns before the proposed scheme is built. An assessment of existing patterns of production on the wet land would be used as a guide to possible future land use without technical change on the scheme land.

As the person responsible for the surveys your first concern would be to specify the objectives more precisely and with an indication of required timing and priorities. At present it is the middle of the dry season. Planning is due to start about one year from now. There is time to do a household survey during the next rains, but the cropping of seasonally inundated land occurs during the dry season, so if specific information is required it must be collected quickly or be deferred until next year and possibly delay planning. The head of the planning department is not very helpful about the precision requirements of the other data. The population information must apply to the whole area and be 'as accurate as possible'. The household information is to be collected from only those communities within a 5 km radius of the proposed scheme. Fortunately, there are two ethnic groups, one of which is known to specialise in vegetable production on the wet lands and grow only a small area of cereals. They also keep small livestock, sheep and goats. The other group farms maize, sorghum and millet extensively, together with cowpeas and groundnuts. In terms of accuracy, the main concern is to identify differences in farm production with sufficient detail to model farm budgets.

In your survey unit you have ten field workers. Two are experienced supervisors with over eight years in the unit. Five of the rest have three years experience and the remainder have just joined the unit as new recruits. Four of your enumerators are women. There is a small group of other professionals in the department, including an agronomist, which you can draw on for short periods. You have two four-wheel drive vehicles and one motor cycle at your disposal.

After considering the problems and reviewing existing data and reports about the study area you decide on the following course of action:

Population data

In order to get a full statement of the settlements and population, you start with aerial photography. The latest available set is five years old, but should give a good guide to settlement distribution. These are mapped on a 1:50 000 scale map. The same photographs are later used by the land resources department to assess current land use patterns.

Having mapped the settlements, the next step is to visit each one and conduct a settlement listing. But there are too many settlements for enumerators to be able to count the population of each. Instead you adopt a three-stage plan. First, in order to obtain an estimate of the number of households in each settlement you interview the community leaders and ask them how many households live in each settlement. Second, you sample a number of settlements and then conduct a household census by visiting and counting every household. Third, when the household survey is done you record the number of people in each of the sample households. The average household size can be used to obtain settlement population (by multiplying average size by the number of households), and the counts of households made at the second stage can be used to correct the estimates of total households made by community leaders at the first stage.

The population data are therefore collected by a combination of census and sample. The exercise is estimated to take between three weeks and one month and can be started immediately. In addition to population numbers, the village listing and household listing forms include provision to record which ethnic group lives in each village. The village listing also includes a provision to record new, abandoned and changed settlements, which may have arisen since the aerial photography was last taken.

Household survey

The household survey is more of a problem. There are no clear guidelines on precision and the complement of field staff would not permit a large study. This will affect the number of settlements to be sampled, the sample size of households within each settlement and the techniques for data collection. Data are also needed from both ethnic groups, who are known to live in separate villages. You decide on a multi-visit survey during the next cropping season, by a mobile team of one supervisor and five enumerators. This will leave some spare enumeration resources in case of accident or illness.

Being aware of the problems with cluster sampling of rural communities you decide on a sample scheme with a large number of clusters and just a few respondents in each. The design you choose is as follows. First, the survey universe is defined in terms of settlements from the village and household listing. Villages are then stratified into two lists representing the two ethnic groups. Within each stratum, villages are selected by simple random sample, and within each village, households are selected by simple random sample at a constant sampling fraction (discussed in Chapter 3). From experience you know that each enumerator can cope with 16 households, as long as crops do not have to be harvested. This would mean a total of 80 households divided between the two strata.

The results would be presented separately for each stratum. At a typical coefficient of variation of 40%, with a sample size of 40 observations the standard error would be about 6% of the mean, which you decide would be acceptable for a planning study.

Data collection would be by a combination of interview and measurement. Farmers would be visited a number of times during the season and asked about crop production activities. Members of the household would be interviewed once to record off-farm activities. Fields and plots would be measured by a tape and compass survey and plot output would be recorded by the farmer's own estimate immediately after harvest.

Wet land study

The wet land study requires a different solution. First, there is great pressure of time, and second, the information required is very detailed, concerning the cropping pattern, rotations, planting decisions, crop husbandry and marketing and consumption. You decide to undertake a rapid reconnaissance survey. You allocate one supervisor plus one enumerator, and co-opt the agronomist from the planning unit to be the technical expert in the study. If he judges it to be necessary he can call on other experts in soils, plant pathology, pest control and marketing, from a nearby research station.

Together you draw up a detailed checklist of topics, and prepare open-ended response forms for the field. Because the interest is in the wet land fields no attempt is made at a formal sampling scheme. Instead, the local extension officer and senior farmers from the community are asked to guide the team to areas of importance. The study is started immediately, and completed when the crops are harvested and before the household survey starts. Equipment is provided for weighing crop output if necessary.

Prices

Price statistics are collected routinely by the national statistical office, although there is a suspicion of variable data quality. You decide to allocate your second supervisor plus one or two enumerators to a monthly check of one or two key markets for the retail prices and black-market prices of crop inputs.

This example illustrates well the way in which different techniques can be combined within the same overall study framework. In fact, a useful semantic difference is between the study and the survey. It is helpful to consider the study as the complete exercise leading to a final report, being made up of surveys and other data collection. In this example, complementary use is made of aerial photography, a sample survey of settlements, a census of households, and a rapid investigation of the wet land agronomy. No single technique could cover all aspects of the study objectives, and even the methods chosen have to be tailored to fit the requirements of enumeration from a resource point of view.

2. *Credit impact*

The general manager of an agricultural development bank wishes to investigate the impact of a new line of credit introduced two years ago. Under this scheme, loans are given to cooperative societies to be on-lent to farmers for the purchase of draught oxen for tillage. Cultivation is done by pairs of animals. The lending programme is supported by one of the major agricultural development funds and was planned under a general credit support project. The oxen loans have been far more successful than was expected. Disbursements rose to over four times the target during the first year, and in the second year exceeded the target for the five-year project. Lending is concentrated at two branches although seven of the nine branches have given loans. The status of loan repayments is not entirely clear as many loans are still in their first year of the four-year term. But repayments at the two major lending branches, from loans in year one, are just 27% of monies due.

As an economist in the planning and research unit of the bank, you are asked to undertake a study to investigate the loans. Specifically, you are asked to examine:

- the source of loan oxen and use made of them

- the production impact of the loans (do oxen-owning farmers produce more than non-owners?)
- the ability of farmers to repay their loans

Within the unit there are five economists and statisticians who can be used for the study. They all have experience of data collection from rural communities.

Agriculturalists within the bank brief you that the main benefits from the loans are expected to come from crop yield increases arising from more thorough and timely cultivation which ownership enables. Non-owners have to hire draught animals and inevitably must wait until owners have completed their own cultivation. This is plausible but careful reading of the project document reveals that in the planning budgets, benefits were assumed to arise from increases in crop area at constant yields. Clearly there is a contradiction to be explored.

Faced with this fundamental inconsistency you decide on a two-part strategy. First you hold a series of exploratory meetings with three groups of people:

- agriculturalists within the bank who manage the loan programme
- branch officers who supervise loans to cooperatives
- three cooperatives from the earliest group of borrowers

The meetings with bank staff are designed to identify the issues involved and prepare a checklist of topics for the main survey. The checklist is then tried out at the three cooperatives. Background information is also collected by contacting research workers and exploring the literature on ox cultivation. Basic loan statistics on disbursement and repayment are requested from all branches for all oxen loans.

Careful review of the first-stage information enables you to draw up priorities for the study. Firstly, you decide that reports from the bank's vet (who has to visit all borrowers to eartag the animals) and branch loan supervision visits will provide sufficient data about where loan oxen were purchased from and their post-loan history. Secondly, it is clear that the main issue for which data are incomplete is farmers' ability to pay. In other words, an accurate assessment of their farm budgets. The general manager wanted a survey covering all lending areas, but enumeration resources are limited. You decide to prepare a design for a case study and to try for a high standard of data quality on a carefully chosen sample. The aspects you concentrate on are how to select farmers for the case study, how to go about collecting production information, and the sample size itself, of course.

The objective for this part of the study is to compare the performance of farmers who own oxen with those who do not. In order to do this you decide to stratify your case study sample into farmers with zero oxen, farmers with one ox, farmers with two oxen, and farmers who used to own zero or one, but now have a loan and own one or two animals. By defining strata with constant oxen resources, seasonal changes from year to year can be distinguished from performance which may be related to oxen ownership. Results are wanted quickly so you cannot wait for two seasons to compare performance before and after a loan. Instead, you decide to collect crop production details by memory recall from the previous year and by farmer estimate for the current year. To strengthen the current year estimates you design the survey to be completed at farmers' fields at harvest time, and arrange for specimen measurement of field areas and local harvest units to correct farmers' area and output measures to metric units.

Even though it is a case study, you decide to make a random sample of cooperatives and farmers within each stratum, to minimise selection bias. Results cannot be generalised to all borrowers, only to the populations used for stratification, but random selection of cooperatives and farmers will add credibility to your findings, and ward off criticism about the findings being peculiar to the sample. Lastly, after consideration of potential workloads, availability of transport and duration of the harvest period you decide that each enumerator could manage sixteen farmers. You allocate each enumerator to two cooperatives, with a sample of two farmers from each of the four strata. For administrative reasons resident enumeration is not possible. The survey is organised for two visits. The first is timed for mid-season to select farmers, enquire about the previous season and prepare for the coming harvest. The second visit includes area and output measurement and lasts over the harvest month.

As a final check of the data, after preliminary tabulations are prepared, you return to every cooperative to check points of details and apparent inconsistencies in the results.

This second example differs from the first. There is not the need for a wide variety of techniques to be employed, but rather a carefully designed study which can concentrate on the central issues. A general agro-economic survey would not necessarily permit differences between owners and non-owners to be analysed. With so much natural and seasonal variation in agricultural data, high standards of accuracy are more important than high statistical precision in this case where budget comparisons are to be made, so a small case study is preferred to a large sample spread over all borrowers. It is implicit in the design that credible accuracy on a small sample will be more valuable than uncertain accuracy on a larger, more representative sample. But within the case study strata, farmers are selected by random sample, to guard against selection bias. Possibly the most important part of the design is the contact between the surveyor and other colleagues and farmers. Planned meetings before the survey are used to define the survey topics and brief the surveyor on the issues. A post-survey visit is used to query points of interest. Notes from these meetings would be used to illustrate the quantitative findings from the survey forms in the study report.

3. *Crop extension*

As regional agricultural officer you are responsible for a programme of crop extension to promote a new pest control and fertiliser regime for rainfed rice. In your region you have thirty extension agents working under a modified version of the World Bank Training and Visit system, promoting technology which was developed by the national research centre, and tested at one site in your area. You want information about the impact of your programme which can be used to guide seasonal management and calculate potential benefits. You have on your staff a monitoring officer whom you ask to design a survey programme to provide information.

After discussions about the objectives of the survey two contrasting needs are identified. Firstly, regular, rapid feedback of farmers' response to extension messages. Secondly, reliable estimates of crop yields at on-farm demonstration sites at the end of the season. There are five enumerators available for data collection.

The two sets of information differ considerably. The extension programme is managed at a district level, and involves a series of technical messages during the year. This means that a survey sample must be large enough to provide results at the district level, and must be synchronised with the technical messages. But the yield study requires accurate and detailed measurement, and will occur once at the end of the season.

Two surveys are designed. The first is a large two-stage (village and farmer) random sample, stratified into contact and non-contact farmers. (Contact farmers are those who are visited regularly by extension agents.) The purpose of the survey is strictly programme monitoring, so the questionnaire is devoted only to extension messages, and asks if the farmer has been given advice, if he has followed it, and his perception of its relevance. The questionnaire is divided into sections according to extension messages, and each one is enumerated within the two-week period following the message. Because each section is short, results can be summarised in simple manual tabulations and circulated to district managers within two weeks of the field visit. Sample size is calculated to give a known precision, which would permit calculation of differences in adoption between contact and non-contact farmers. (For an example designed for agricultural extension see Murphy and Marchant (1988)).

The yield study will be used as the culmination of the demonstration process with contact farmers, as well as to document the performance of the crop packages. Each extension agent has eight demonstration plots. You decide on a total harvest of each plot. To implement the survey you assign the five enumerators to supervise the extension workers harvest their plots, after first preparing a manual covering harvest techniques and holding a training session for all the field workers. As the harvest is partly a demonstration exercise, the contact farmers are also briefed and encouraged to help in the work and crop weighing. To ensure that the yields are reported accurately, samples of crop are measured for moisture content and the harvested area is measured by a tape and compass survey.

In this example the strategy has been to support programme management by surveys which are narrow in their focus of topics and timed to coincide with management activities. Random samples are important to generalise farmer adoption, but the harvest is restricted to the demonstration plots. By documenting the harvest procedure and training staff you try to ensure comparability between crop locations.

1.4 Some general guidelines

The discussion has stressed the need for flexibility in the planning of investigations, and the importance of letting the objectives of the study, and the constraints of the available resources, determine the methods to be used. Preconceptions, prejudices, or a blind following of precedent, should not influence this process.

When considering the choice of the type of enquiry or investigation to be conducted, we have stressed the independence of the three dimensions of design, scale and method. It should never be assumed that a particular type of enquiry inevitably implies a particular choice of data collection method, or a specific scale of enquiry implies a specific design. Each aspect should be considered on an individual and independent basis – though some combinations do fit together better than others. A further point is that one study may be best conducted by a combination of several different enquiries, aimed perhaps at distinct topics, or stages of the investigation with different levels of detail or precision needed.

But there are still likely to be several possible approaches to obtaining information on any particular topic, and then a choice must be made between them. Here the authors take a definite stand. Within the limits of possibility and practicability, *the more formal and structured an approach the better*. This is because information obtained by a more formal approach can be confidently generalised and is more widely applicable and comparable between different locations, enumerators and points in time. If there is a choice between a random and a non-random sample, then the random sample is to be preferred because it produces generalisable data, and it is likely to be worth expending time and resources on producing a sampling frame in order to achieve the greater efficiency of a random sample (see Scott, 1985, Chapter 3). If there is a choice between a structured and an unstructured interview, then some structure, whether using a checklist, a questionnaire, or some other device, is preferable because it ensures consistent and comparable coverage of the topics of interest. And it is worthwhile to choose the more formal option even where only a part of a survey design is involved, for instance in taking a random subsample of farmers or households after taking a non-random sample of villages.

Notes

1. The use of low-level aerial survey techniques for resource management is reviewed in ILCA, 1981.
2. See, for example Salmen (1987), Carruthers and Chambers (1981) and Hildebrand (1981).
3. The problems of experimentation and the inference of causality in the context of agricultural surveys are discussed in Casley and Kumar (1987), Chapter 8.

2

Introduction to sampling: ideas and considerations

2.1 Basic concepts of sampling

The rationale of sampling

The purpose of sampling is to economise on the use of resources in gathering information. There are many situations where it is impracticable or uneconomic to obtain data on the whole of the population of interest (the universe) and so information is collected from only a part of that population (the sample).

When we speak of a sample survey, we generally mean that the sample in question is a random sample – one chosen entirely by chance, in such a way that every element of the population has a known chance of being selected. This type of sampling normally (but not always) requires a sampling frame, from which the sample can be selected.

In some enquiries, a non-random sample – a purposive, or informal one – may be used. In such a case, the sample is selected not by chance, but for reasons of practical convenience, or in order to include items with particular characteristics in the sample. For example, in an exploratory survey, where a sample of farmers is interviewed concerning the types of crops and crop mixtures they grow, an informal sample may be taken simply by interviewing the farmers which the investigating team happens to meet while travelling around the study area. In a case study to assess the effectiveness of a package of new techniques under farmer management, a purposive sample of farmers may be chosen to cover different soil types and climatic conditions, and different sized landholdings. Non-random sampling is most likely to be found in rapid appraisal surveys and in case studies, but it is not necessarily the best approach. Samples selected in whole or in part by random methods may also be found.

The main advantage of random sampling is that it produces generalisable estimates – information obtained from a random sample can safely be generalised to apply to the population as a whole. With a non-random sample, there is no way of telling whether the sample is typical of the whole population. The main disadvantage of random sampling is that it requires sampling frames to be constructed, which increases the costs of this type of sampling, and the time needed. Non-random sampling does not. There is however, often a trade-off here. A random sample is likely to be more costly, but also more effective than a non-random one. It is sometimes better at an early stage to put resources into creating a frame, which will enable the investigator to work with a smaller sample with a known relation to the population than to try to interpret results from a larger sample which cannot confidently be generalised to the population. The trade-off is between the resources which have to be diverted from the main sample (and thus give rise to a smaller sample size) to construction of a sample frame, which permits random selection (but at a lower level of precision than the original sample size) (Scott, 1985: 10). It is not the case that random sampling is always associated with a large-scale survey, and non-random sampling with a small-scale one. Nor is random sampling invariably found together with a formal questionnaire, and informal sampling with an unstructured interview. It is not even the case that any one survey design must involve either random or non-random sampling, but not both. A design involving non-random sampling at one stage, and random sub-sampling at a later stage, is quite possible, and may be particularly suitable for certain kinds of studies, as is further discussed in Chapter 3.

These points notwithstanding, there are certain circumstances under which either random or non-random sampling is more suitable. Non-random sampling is likely to be more appropriate when information is needed very rapidly, but not with great precision, for example in an initial, exploratory investigation. It may also be appropriate for very detailed studies, for example in testing research station findings under ordinary farm conditions. Where a detailed study must also cover a range of cases from different sub-populations, such as farms in different ecological areas or altitude bands, then purposive sampling is particularly likely to be found. Random sampling is especially appropriate in the case of a large-scale survey, with simple questions, where the results must be generalised to the whole population.

Given the focus of this book on formal survey methods, however, it is random or scientific sampling techniques which will be discussed in this and the next chapter.

After a sample has been selected and data collected from it, these sample data are used to make estimates of the characteristics of the whole population. To do this, an estimator is used – a formula for calculating a population estimate in terms of the sample observations. The appropriate form of estimator, so that a valid inference can be made from the sample to the population, depends on the sample design used. If the design is one where every unit of the population has an equal chance of being selected for the sample then we have a particularly convenient situation in which the information from the sample units can be very simply combined to obtain estimates of the characteristics of the whole population. If the units have different probabilities of selection, then it is necessary to 'weight' the information from them, that is to multiply the data from the different sample households by different factors when combining them. The question of weighting, and derivation of estimates, is discussed in more detail later.

To summarise, the design of a formal sample survey must therefore ensure that a proper random sample is obtained and that the calculation of estimates from it is done correctly, including any necessary weighting, so that accurate and generalisable estimates are obtained. It is also desirable to design the sample as efficiently as possible, so that the available resources are used effectively to produce the required information.

Sampling errors

When we say that a random sample is representative of the population which the sample is selected from, and that this gives results which are generalisable to that population, this does not mean that an estimate of some characteristic obtained from the sample will necessarily be exactly the same as the true value of that characteristic in the whole population. There is quite likely to be a difference between the estimate and the true population value, and a part at least of that difference (other possible differences are discussed below) is simply due to the estimate coming from a part of the population rather than the whole of it. This element of difference between the estimate and the true value is known as the sampling error. The word 'error' in this expression is not used in the sense of a mistake or fault – it is the inevitable consequence of sampling and represents the uncertainty introduced by sampling.

If repeated samples were taken from the same population and estimates obtained from each, all the estimates would be affected by sampling error, and a number of different values would be obtained. The range of sample values form a sampling distribution of an estimator. An example will help to understand this concept.[1]

Suppose we have a large population of extension contact farmers, who have had a total of four new cultivation practices recommended to them by the extension agents. We wish to know how many, on average, of these recommended practices have been adopted by the farmers. Let us consider for simplicity a situation where the adoption of each possible number of practices (0, 1, 2, 3, or 4) is equally likely. Thus the true population average or mean m is given by

$$m = \frac{1}{N}\Sigma X_i \text{ for } i = 1,2, \ldots N$$

where X_i refers to individual observations of the number of practices adopted and N is the total number of observations.

Because the adoption of each possible number of practices is assumed equally likely the average adoption m, is calculated by

$$m = \frac{0 + 1 + 2 + 3 + 4}{5} = 2$$

Now consider drawing a sample of just two farmers (sample number $n = 2$). What is the range of possible results for the number of practices adopted? All the possible combinations can be set out in a two-way table (Table 2.1). A total of 25 samples would be needed to cover all possible

Table 2.1. *Combination of sample observations*

		Second farmer Number of practices adopted				
		0	1	2	3	4
First farmer	0	0:0	0:1	0:2	0:3	0:4
	1	1:0	1:1	1:2	1:3	1:4
	2	2:0	2:1	2:2	2:3	2:4
	3	3:0	3:1	3:2	3:3	3:4
	4	4:0	4:1	4:2	4:3	4:4

combinations of adoption by the two farmers. The table shows the number of practices adopted by the first farmer combined with successive numbers of practices by the second farmer. The next step is to sum the values in each cell of the table and divide by two to calculate the average number of practices adopted for each sample ($\bar{x}$) (Table 2.2). Now we can count how many times each value of the sample mean appears, and list it in a frequency distribution table (Table 2.3).

This is the sampling distribution of the mean for a sample of two farmers. From the frequencies we can see that a sample mean of two, equal to the population mean, occurs more often than any other value. Even so, it only occurs five times out of 25. Although the sample mean only equals the population mean in 20% of the samples, there is a tendency for the other values to be close to the mean. Nineteen of the 25

Table 2.2. *Sample means* ($\bar{x}$)

		Second farmer Number of practices adopted				
		0	1	2	3	4
First farmer	0	0	0.5	1	1.5	2
	1	0.5	1	1.5	2	2.5
	2	1	1.5	2	2.5	3
	3	1.5	2	2.5	3	3.5
	4	2	2.5	3	3.5	4

Table 2.3. *Frequency distribution*

Sample mean ($\bar{x}$)	Number of samples
0.0	1
0.5	2
1.0	3
1.5	4
2.0	5
2.5	4
3.0	3
3.5	2
4.0	1
Total	25

samples (76%) lie between values of one and three. Lastly, if we sum all the sample means in the table, and divide by 25, to calculate the mean of the sampling distribution $\bar{\bar{x}}$, it will be found to equal two, the population mean. Thus from our example we can see that there is a tendency for sample means to be clustered around the population mean; that more sample means will be close to the population mean than far from it; and that the mean of the sampling distribution equals the population mean.

The shape of the frequency distribution is seen from a histogram of sample frequencies (Figure 2.1). The shape is peaked and symmetrical about the mean. If we were to repeat the exercise with samples of three, four or five farmers, the frequency distribution would start to approximate to a normal distribution.[2] This is an important finding. In this example, the distribution of the population values is not normal, but was defined by equal probabilities of occurrence for each value. Yet despite this the distribution of sample means is approximately normal. It is this important feature of the distribution of sample means, known as the Central Limit Theorem, which enables population estimates to be made from sample values. Practically, this is a fortunate situation for samplers as population data do not have to be distributed normally for statistical inferences to be valid (Caswell, 1989:219. Rees, 1985:72).

In this example we are able to draw every possible sample of size $n = 2$, and the frequency distribution of those means showed all possible outcomes and their associated frequency. But in a practical sampling situation we would have only one observation of a sample mean, yet we would still like to know, at a certain level of confidence, how close it may

Figure 2.1.

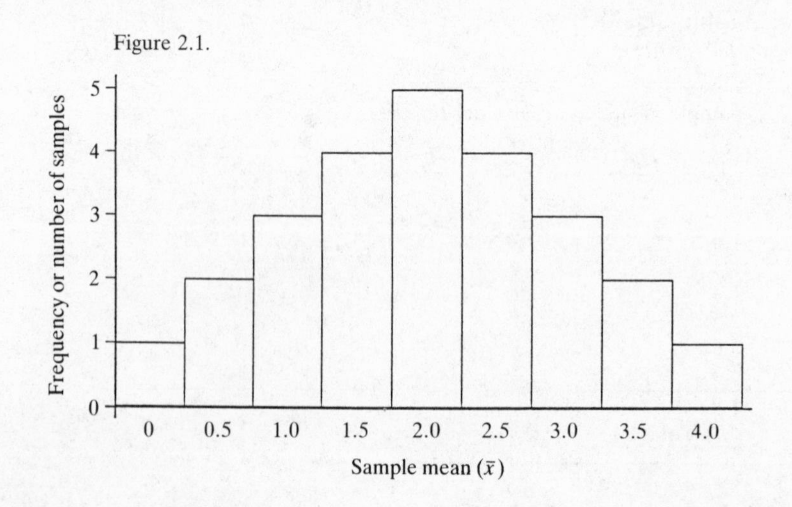

be to the true population mean. We can find this out by using the properties of the normal distribution.

Figure 2.2 shows two normal distributions with different shapes. The first is a peaked distribution, with the observations close to the mean. The second is more widely spread. The link between the two is that the area under the curve has a constant relationship with the standard deviation of the distribution.[3]

The standard deviation is calculated as the square root of the variance. For a population of size N and individual observations X

$$\text{SD}(X) = \sqrt{\text{Var}(X)} = \sqrt{\left(\frac{\Sigma(X_i - m)^2}{N}\right)}$$

In a normal distribution, 68% of the area under the curve (that is to say 68% of the observations) fall in the range of ±1 standard deviation either side of the mean. A further 27% (totalling 95%) are enclosed by approximately two standard deviations, with the remainder falling outside this range.

Figure 2.2.

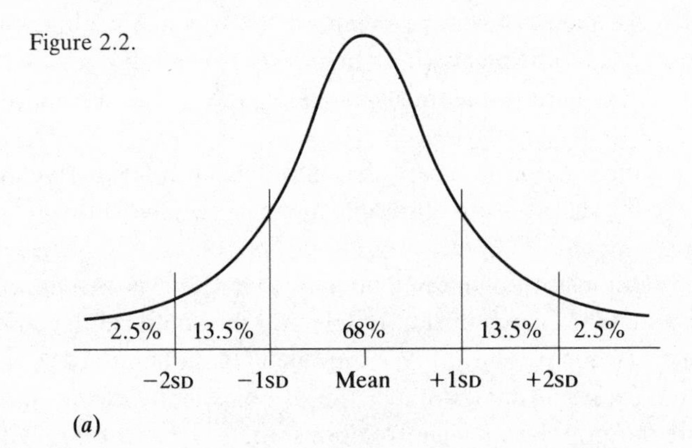

(*a*)

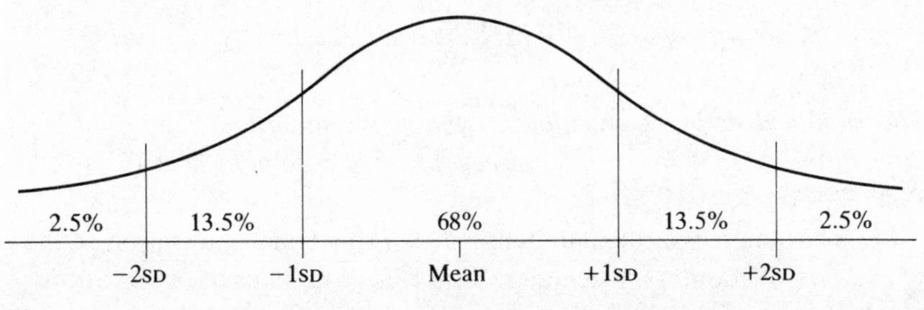

(*b*)

A peaked, clustered distribution will have a standard deviation which is small in comparison to the mean. Expressed as a percentage ($\text{SD}/\bar{x}\%$), this may be of the order of 5 to 10%. A more widely spread distribution will have a large standard deviation, perhaps 40% or more of the mean.

Returning to our example of adopters, as we know all the possible sample means ($\bar{x}$) we can calculate the standard deviation of sampling means about the mean of the sample distribution ($\bar{\bar{x}} = 2$).

$$\text{SD}(\bar{x}) = \sqrt{\text{Var}(\bar{x})} = \sqrt{\left(\frac{\Sigma(\bar{x} - \bar{\bar{x}})^2}{N}\right)}$$

Working from the data, where $\bar{\bar{x}} = 2$ and $N = 25$ the standard deviation of sample means can be calculated as 1.0.

At this point we must introduce some new terminology. The standard deviation of sample means is called the *standard error*. This term is used from here on.

Knowing that in our sample the standard error has a value of 1.0, and using the properties of the normal distribution, we can say that there is a 68% chance that the sample mean would be located in a range of 1.0 unit either side of the mean of sample means ($\bar{\bar{x}} = 2$), which we know to be equal to the population mean (m). In stating this we have attached a probability to the sample mean, using the shape of the normal distribution.

This is extremely helpful, except that in practical situations we do not have access to the distribution of sample means – we have only one value for our sample mean.

It can be demonstrated mathematically that there is a relationship between the standard error of the sample and the standard deviation of the population (when N is large). We can calculate the standard deviation of our sample from our data, and use this as an estimate of the standard deviation of the population. The relationship, which we offer without proof, is stated as

$$\text{Standard error} = \text{SE}(X) = \frac{\text{SD}(X)}{\sqrt{n}}$$

The standard error is calculated from the standard deviation of the population divided by the square root of the sample size (Caswell, 1989:220. Rees, 1985:75).

In our example we calculated $\text{SE} = 1.0$. The standard deviation of the population of adopters, calculated from a large number of observations, was 1.414, and the sample size (n) was 2, thus:

$$\text{SE} = 1 = \frac{1.414}{\sqrt{2}}$$

which is correct. Using this relationship, we can now see how SE would change with larger sample sizes.

$$n = 5 \qquad \frac{1.414}{\sqrt{5}} = 0.63$$

$$n = 10 \qquad \frac{1.414}{\sqrt{10}} = 0.45$$

$$n = 50 \qquad \frac{1.414}{\sqrt{50}} = 0.20$$

Obviously, the standard error gets smaller with increasing sample size, but notice that it reduces with the square root of sample size. Thus to halve the standard error, sample size must be increased fourfold. This relationship forms the basis for the estimation of sample size and is discussed further in Chapter 3.

Confidence intervals

Having presented the basic concepts of sampling and estimation procedure, we can now explain how an estimate of the population mean can be expressed in probability terms. From the distribution of the mean of sample means (m) we have seen that 68% of the values of the sample mean ($\bar{x}$) lie within one standard error of the mean of sample means. This can be written as a probability P.

$$P[m - 1\text{SE} \leq \bar{x} \leq m + 1\text{SE}] = 0.68$$

This can be rearranged to state

$$P[\bar{x} - 1\text{SE} \leq m \leq \bar{x} + 1\text{SE}] = 0.68$$

The value of 68% is an uncommon probability to use as it represents a low chance of around 2:1. It is more usual to work with a level of 95%, equivalent to 19:1 and equal to 1.96 standard errors.

$$P[\bar{x} - 1.96\text{SE} \leq m \leq \bar{x} + 1.96\text{SE}] = 0.95$$

The interval $\bar{x} - 1.96\text{SE}$ to $\bar{x} + 1.96\text{SE}$ is the 95% confidence interval for the population mean. It is interpreted as follows: there is a probability of 95% (or a chance of 19 in 20) that the calculated interval contains the

Table 2.4.

Probability (%)	z
68	1.00
75	1.15
80	1.28
85	1.44
90	1.64
95	1.96
99	2.58

unknown population mean. The chance that the mean is not in this range is equivalent to 1 in 20.

Probabilities can be calculated at any chosen level, using statistical tables of the proportion of the area under the normal curve, usually referred to as z values. Some selected values are shown in Table 2.4.

z values are used for large samples where $n \geq 30$. The distribution of means for small samples follows the 't' distribution and the calculations need to be modified accordingly (Caswell, 1989:115; 220. Rees, 1985:63; 80).

To summarise, the standard error is a measure of how variable the estimates from different samples would be and hence how close the estimate from a single sample is likely to be to the true population value. From a random sample, it is possible both to estimate a characteristic, and also to estimate to what extent it is likely to be affected by sampling error.

Summary

Sampling error

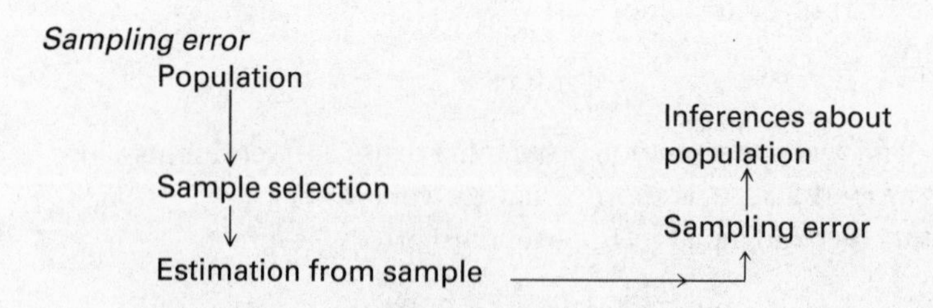

Non-sampling error

Sampling error occurs because the estimate is from a sample rather than from the whole universe. There is also non-sampling error, which is a general expression covering all kinds of errors which are not related to sampling and which would arise even if the whole universe was studied. Some common types of non-sampling error are:

- response errors
- measurement errors
- recording and transcription errors
- processing errors
- selection bias

Unlike the sampling error, the non-sampling error of an estimate cannot itself be estimated, except in special circumstances. And yet the non-sampling error is likely to be at least as important as the sampling error (Casley and Kumar, 1988:84).

It is very often the case that measures which may reduce sampling errors may at the same time increase non-sampling errors. This is particularly clear and striking when considering the question of sample size. Generally speaking an increase in the sample size for a survey reduces sampling error – but if at the same time it increases the workload on field and supervisory staff, or demands a large work force which is liable to be of lower quality overall, the quality of the data collection and processing work is likely to deteriorate and the non-sampling errors to increase. This reciprocal relationship should always be borne in mind when designing a survey.

One important point is that if the sample that is drawn is biased in some way – due to an incomplete frame or other exclusion of part of the population, the resulting selection bias would be a component of non-sampling error. The effect would be that the mean of the sampling distribution would not equal the true population mean – in other words, the estimator is biased. Sampling error, by contrast, is the difference between the sample mean and the mean of the sampling distribution.

Accuracy, precision, and bias

These are all words which are found in ordinary, common usage, but which are also used in the context of sampling theory, where they have precise, technical meanings. In sampling theory the accuracy of a sample estimate is the measure of its difference from the true population value. It is usually rather easier, however, to assess its difference from the mean of the sampling distribution, this is known as its precision and is measured by the standard error. The difference between the mean of the sampling distribution and the true population value is the bias. The three concepts are thus related as in Figure 2.3.

Consider, for example, an investigator who wishes to know the price of tomatoes in a market, and records the price paid for a weighed quantity from each of a sample of sellers of tomatoes. If the variations in price between different sellers is wide, and/or only a few are sampled, then the precision of the estimate obtained is likely to be poor, because the sample values can be expected to vary widely about their mean. If the scales are not true, but read too high or too low, then the estimate will be inaccurate. Inaccurate samples will bias the results, as also would a sampling scheme which favours particular types of sellers according to factors such as location, time of selling, etc. A biased estimate means that the mean of the sampling distribution differs from the true value.

Earlier in this chapter the mean of the sampling distribution was found to be equal to the true population mean. This estimator was thus unbiased, ie. its bias was zero. Most estimators in common use are unbiased, though occasionally it may be convenient to use an estimator which suffers from some small degree of bias, because it is otherwise particularly suitable and convenient. This kind of bias, which is a characteristic of the estimator, is sometimes known as sampling bias, but a much more common and usual source of bias is non-sampling error. For example, if non-response to a question on income is particularly high among high-income households, then any estimate of income for the whole population will be biased. If a crop harvest is weighed using scales which have not been properly calibrated and give a reading consistently higher or lower than the true value, then the estimates of crop harvest will be biased. If an enumerator stationed in one village is better at obtaining true information about the extent of farmers' landholdings than an enumerator in another village, any estimate of the difference in size of holdings between the two villages will be biased.

Figure 2.3.

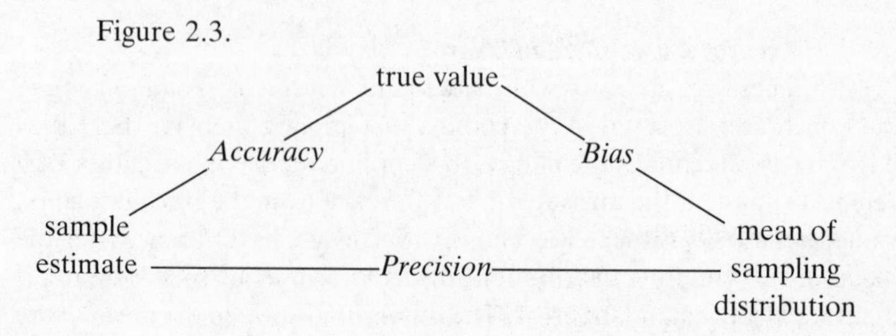

It is not usually possible to measure the extent of bias, because it requires a knowledge of the true population value. The precision of an estimate, however, relates to sampling error only, and is measured by means of its standard error. Precision is thus the easiest of the three concepts to measure and so the most commonly reported. It is useful and important, but seldom conveys the full extent of survey error.

A useful, and commonly-used, criterion for comparing estimates which may have different amounts of bias is the mean square error (MSE), or its square root, the root mean square error (RMSE). The MSE is defined as the mean of the squares of the deviations of the estimates obtained from all possible samples, from the true population value. It can more easily be thought of as the sum of the variance of the estimate (the square of the standard error) and the square of the bias – it thus combines the effects of sampling error and bias.

If a non-sampling error leads to a relatively large bias then this factor is likely to dominate the RMSE. For example, if the uncertainty due to non-sampling error is measured by a standard deviation of 10% of the mean, and the sampling error is 5% of the mean, then the RMSE r is given by:

$$r = \sqrt{(0.10^2 + 0.05^2)}$$
$$r = 0.112$$

The total error, r, is 11.2%. If the sampling error were to be halved to 2.5% (implying a fourfold increase in sample size, see Chapter 3) the RMSE is reduced only to 10.3%. If the increase in sample size led to a reduction in data quality and an increase in non-sampling error, the component of non-sampling error would dominate total error and lead to an increase in RMSE (Casley and Lury, 1981: 86).

2.2 Sampling frames

An understanding of basic sampling concepts, as discussed above, is important for sample design and data collection. However, equally important are a number of practical considerations – at design stage – which affect the success with which sampling procedure is carried out.

Sampling units

In planning a sample survey, one of the earliest decisions must be about what is to be the sampling unit – the type of unit which is to be selected, and interviewed or measured. Sometimes the choice of sampling unit is obvious, but often it is a more complex issue than is apparent at first sight. In a survey to examine farm cropping patterns, for example, is the sampling unit to be the farm, the farmer, the parcel of land or the individual plot of cultivation? The primary issue to be determined is within which unit are inter-relationships to be examined? This consideration may require for example that the unit be the farm household rather than the farmer, where economic matters are being investigated, or the farm rather than the plot, where costs are the subject of the survey. But if, for example, the desired information relates to inputs for a specific crop, it may in practice be most convenient to treat the whole farm as the sampling unit, and to extract the information relating to the crop of interest at a later stage.

Definition of universe

Another early decision which has to be made is in defining the universe of study, the population which the sample is to represent. This might be, for example, all farming households in a country, or all farms receiving credit under a particular scheme, or all herds of cattle in a specific region. It is important to be precise in the definitions used – in particular, the specification of what constitutes a farming household, or a livestock-owning household, can often cause problems unless the issue is carefully addressed beforehand. There are likely to be households which are only part-time farming ones, perhaps cultivating a little land for family subsistence or keeping a few chickens. Are they to be included in the universe? What is the minimum level of farming activity which is relevant for this survey?

It is also necessary to decide what to do about sub-groups of the population which are particularly difficult to obtain information about, or unusual in some way, and to state explicitly whether they are to be included in the survey universe. Examples of such sub-groups might be farmers living in a particular sparsely populated or almost inaccessible area; people living in institutions rather than ordinary households; or part-time and absentee farmers.

It is helpful to write down as exactly as possible a definition of the survey universe, and then to try to think of 'borderline' or awkward cases, decide whether they should be included, and if necessary revise the definition according to those decisions. In a situation where the survey is being directed towards a specifically defined population, the term target population is sometimes used.

Accuracy of frames

Sampling frames can be of a number of different kinds. The most common is a list, often especially constructed, for example a list of all the households in a village obtained by an enumerator going to each dwelling in turn and writing down the name of the head(s) of household living there. Other kinds of lists can be obtained from already existing sources, for example a card index belonging to an agricultural cooperative containing the details of all farmers belonging to it, or a computerised database held by a bank containing information on farmers who have obtained loans from it. A map is also a common element in a sampling frame, particularly for agricultural topics – for example, a map showing the locations of all the plots of cultivation receiving water from an irrigation scheme. There are often several stages of sampling for a survey, each with its own frame, and these may be of different types at the different stages.

A frame may be procured or adapted from another source, or it may be constructed especially for the survey, but in either case it is critical to ensure that it is accurate. First of all, the frame must cover precisely the survey universe. This requires that the geographical coverage, and the definitions of who or what are to be included in the universe, must be clearly and explicitly defined, and furthermore that it must be possible to observe them in practice.

The next important point is to ensure that, within the universe of study, there are no omissions or duplications in the sampling frame. These are most likely to occur where the boundary between different portions of the frame is not clearly defined. For example, if two field workers engaged in listing households for a frame are given sketch maps showing the geographical area each is to cover, and the exact physical boundary between their areas on the ground is not clear, then households in the general area of that boundary may be listed by either, both, or neither. Another common omission is when households are listed with reference to the dwelling units they occupy, but instructions to call back to dwelling units which appear empty or produce no response have not been given to the field workers.

A further possible flaw in a sampling frame is that of simple inaccuracy. For instance, in a map-based frame, a village may be located in the wrong place, or in a list of households the name of the household head may be incorrectly recorded.

All these errors may arise as a result of inadequate updating of the frame. It is vital to check whether it is up to date, especially where a frame is taken from another source. If necessary, an updating operation should be carried out. Even when a frame has been specially created for a survey, it is all too easy to do so too early so that it becomes outdated, or to assume that a frame constructed for an earlier survey in the programme is still suitable (Casley and Lury, 1981:72).

Frames for multi-stage designs

There are many instances where a relatively large-scale survey is to be conducted, no suitable frame already exists, and the cost of constructing a complete frame for the survey is prohibitive. A common solution to this problem is to use a multi-stage sampling design (discussed further in Chapter 3). In this case, it is only necessary initially to construct a frame of the larger, first-stage units, and then to construct more detailed frames only for those first-stage units selected, and so on until the final stage is reached of selecting the ultimate sampling units. As an example, a common type of multi-stage design is one where the ultimate sampling unit is to be the household, but at the first stage the unit is the village, or some other convenient locality or settlement type. In the first instance a list of villages is needed, and it is often possible to obtain existing lists from the local administration. From these a sample of villages is selected, and lists of households created for the selected villages only, from which the sample households are chosen.

It is often possible to use a sample from one survey as the frame for another, and this possibility is always worth investigating, since it saves most of the cost of frame preparation. This approach is particularly appropriate where an integrated programme of surveys is being conducted, since it is likely to be useful for analysis and interpretation to conduct some surveys on subsamples of others.

Additional information in sampling frames

There is a number of possible reasons for collecting additional information during the construction of a sampling frame. One, which has been touched on above, occurs when the definition of the universe or of the sampling unit to be covered is rather complicated to apply under field conditions, and so classificatory information is gathered during the frame listing, and the final decisions as to which units are to be excluded or included can be made at a later stage. This situation is illustrated by Example 2.1, where the frame required is of farming households, but the definition of this is rather complicated, so that the initial listing is to be done of all households, and at the same time information is gathered and recorded which will enable those households which are defined as farming households to be identified, for inclusion in the sampling frame.

EXAMPLE 2.1

The form given below is designed for the preliminary listing for a sample frame of farming households. In this case, a farming household is defined to be one where

- **the head of household does farm work full-time; or**
- **where any crops were sold during the last agricultural season; or**
- **where at least two head of cattle or five sheep or goats are owned.**

All households are listed, together with the information needed to decide whether they are farming households under this definition.

Sampling frame listing
Extension contact survey

Village .... Enumerator ....

Code ☐☐☐ Supervisor ....

House no.	Household no.	Head of household	Whether works farm		Whether sold crops	No. livestock owned		Office use only		
			full-time	part-time		cattle	small stock	if in frame	frame no.	if in sample

---

A similar approach is taken if the frame is to be used for two or more different surveys, where the universes may be different, that for one survey being perhaps a subset of that for another. An example of this situation might be where one survey is to be conducted on all households, and a second survey on farming households only.

A further common reason for the inclusion of additional information in the frame listing is for purposes of stratification (see Chapter 3), where the sampling units are to be divided into several distinct categories, and a sample selected for each category separately. For example, it might be desirable to have separate strata for farming households owning no livestock, owning cattle only, owning small stock only, or owning both. In this situation, the stratifying information must be gathered and recorded during the frame listing. It is usually preferable for the necessary data to be recorded first, and for the strata to be determined later, rather than for the enumerator to designate strata in the field while doing the listing, as this latter approach is more likely to lead to errors, while being at the same time more difficult to check.

Costs of frame preparation

The preparation of sampling frames can be an expensive exercise. At least for the final stage of sampling, it is usually necessary for all the sampling units in the frame to be visited and recorded. And, assuming the sample design is an efficient one, the number of units listed for the frame is very much larger than the number in the selected sample for the survey itself. It is therefore very advisable to reduce the amount of frame listing needed as far as possible by the devices discussed above, such as multi-stage sample designs, devising frames from available information, using frames for several surveys, and using the sample for one survey as the frame for another.

But having reduced the frame preparation work as far as possible, there will usually still be work which needs to be done to create a good sampling frame, and the costs of this work are unavoidable. It is a false economy to try to cut corners on frame preparation by skimping on field work or avoiding difficult areas or issues. A sample can only be as good as its frame, and if the frame is defective then the survey results cannot properly be generalised.

Also, there tends to be a trade-off between the frame preparation and the survey fieldwork itself. A well-constructed sample frame can often save time in the survey by improving the logistics of the fieldwork or by enabling a more efficient sample design to be used, and hence a smaller sample size. Thus to some extent resources may be more effectively utilised in frame preparation than in the survey proper, and it is worth considering whether costs might be better transferred between these survey phases. A suggested rule-of-thumb is given in a World Bank technical publication on monitoring and evaluation:

> If no list is available and if the creation of a list is limited only by cost constraints, then it would be worthwhile to sacrifice one quarter to one third of the planned sample size in order to release funds to carry out the listing (Scott, 1985:10).

Above all, the costs of frame preparation must be considered at the planning stage, and must be budgeted for. They are likely to be a significant proportion of the total costs, and relate to an element of the survey work which is of critical importance in determining the eventual quality of the survey results. It is vital that adequate provision be made for this work.

Summary

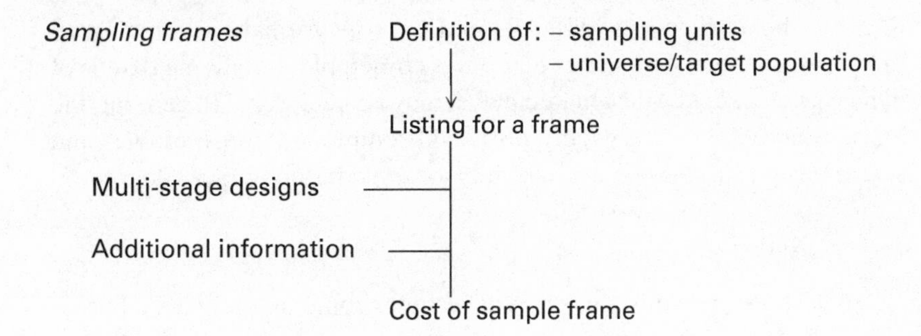

2.3 Survey design implications for field organisation

The type of sample design chosen for a survey has implications for the organisation of fieldwork for the survey. A factor of particular significance is the degree of clustering of the sample units. Interacting with this, is whether the survey is a single-visit one, or whether it requires frequent visits, for multiple interviews or continuous monitoring.

If the survey is a clustered, single-visit, one then it is most conveniently conducted by teams of enumerators travelling together from cluster to cluster. This is a straightforward organisation for supervision, but requires some thought regarding the arrangements for accommodation and transport.

If the survey is a clustered, multiple-visit, type then it is probable that the enumerators will need to be stationed for part of the time in the clusters, and will probably need to be accommodated there. If the clusters are large enough, or the survey detailed enough, to require several enumerators per cluster, then complete teams can be located together, which has advantages for morale and for ease of supervision. This consideration may well conflict, however, with the need to keep the cluster size small in order to reduce intra-cluster correlation effects, and so it may be necessary to locate enumerators individually. In this case, careful thought needs to be given to the supervision arrangements, and to the logistic support for them.

For a sample design which is not geographically clustered, the enumerators will probably need to be individually mobile irrespective of whether the survey is of a single-visit or multiple-visit type. This often requires individual transport facilities (for example bicycles, or motor bikes) to be arranged for enumerators as well as supervisors. There may also be difficulties in arranging accommodation for field staff. Supervision is likely to be particularly difficult, and a fairly formal system of pre-planning and recording the enumerator's timetable, or leaving details of daily movements at a designated place, may be necessary. In general, the logistic difficulties of a widely dispersed sample are considerable, and such a design should be very carefully assessed before being adopted.

Notes

1. The material in this chapter includes some mathematics, but no more than is essential for a grasp of practical sampling. Two basic statistics texts are referenced in this and subsequent chapters for readers who seek an alternative exposition: Rees (1985) and Caswell (1989). Readers who prefer a more mathematical approach are referred to Cochran (1977) and Snedecor and Cochran (1967). All readers would benefit from the commentary in Casley and Lury (1981).
2. A normal distribution has a bell-shaped curve, is peaked, and symmetrical about the mean. See Rees (1985:63) and Caswell (1989:33; 114; 209).
3. The standard deviation is one measure of variation. See further in Chapter 10; Rees (1985:22), and Caswell (1989:100).

3

Sample survey design

3.1 Sample survey statistical designs

It is already apparent that different designs of sample are likely to be appropriate for different types of survey, and in different circumstances. This is a very large subject, and can not be treated comprehensively here – the reader is referred to more specialised texts if a fuller treatment is required; see for instance, Yates (1949), Cochran (1977). We consider here some of the designs likely to be most appropriate for sample surveys concerned with agricultural development, and the considerations which influence the choice of design.

This chapter describes the sampling designs, and the procedures for drawing samples according to these designs. The procedures for estimating means, totals, ratios and proportions, and their standard errors, for these designs, are given in Appendix 1.

Informal sampling

Before considering formal sample designs – those using formalised systems of selection from sample frames – we shall first consider informal sampling. An informal approach to data gathering is often used in exploratory or preliminary investigations, where the construction of a frame is too expensive or too time-consuming, yet some approximate information is needed as quickly as possible. It was argued in Chapter 1 that even in these circumstances it is worthwhile to randomise the selection of informants to whatever extent possible, in order to achieve some degree of generalisation.

One very common informal design is where the investigator or team makes a purposive selection of villages and then interviews a number of informants in each village. Unless households with specific characteristics are required for interview, the selection of informants in each village can be randomised by a simple procedure such as this.

> Starting from the house of the village head walk along the nearest path to an easterly direction. At the first junction take the path to your left. Select the tenth house you come to. From that household turn left, through a right angle (90 degrees) and continue to the next tenth house. This is the second unit in your sample. After that household turn right and continue to the next tenth household. Continue in this way alternating right or left after selecting another household until the required sample has be chosen. If your path starts to take you away from the settlement at any time turn alternately left or right to walk along the perimeter and re-enter the village. If some respondents are absent when you select their household return later in the day.

The starting point, starting direction and number of households counted could all be different from those in the example, depending on local circumstances. The essential points are that the procedure gives people living in different areas of the village roughly equal chances of being interviewed and the interviewer does not concentrate on those living in the centre or those who come forward to be interviewed, or those introduced by the head or by the extension agent or the local government representative.

Another common informal design is to select a number of plots belonging to a village and to examine the plot characteristics (cropping pattern, cultivation practices, etc.) by observation and by interviewing the farmers. It is possible to randomise the selection of plots by a similar procedure to the one for households.

> Start from the village mosque. Choose a random compass bearing from a table of random numbers or by throwing a pointed stick in the air and taking the direction it points on landing, and walk along that bearing. When the beginning of the cultivation is reached start counting paces. After each 250 paces stop, and select the plot in which you are standing. After reaching the outer edge of cultivation choose a second random bearing and repeat the exercise.

Again, the start point and interval between selections would depend on local circumstances. This procedure makes sure that plots at different distances from the village centre are selected and the choice is not concentrated on those close to the settlement, or on those easy to reach by good tracks. The use of a second random bearing will help avoid bias if the first bearing follows a particularly favourable or unfavourable feature such as a stream.

Many such informal sampling designs can be devised for different purposes and in different circumstances. The most important considerations are to avoid the biases which are likely to occur if informants are selected according to some person's preferences. The chosen informants should therefore be spread around the study area or scattered throughout the study population. In this way, results from the survey will be more plausibly true for the population as a whole (Casley and Kumar, 1988:78).

Simple random sample (SRS)

Turning now to formal designs, the simplest kind of sample design is the appropriately-named Simple Random Sample (SRS). This design is often used to select a sample of households or farmers from all those in a particular area, or to choose a sample of villages or other localities. It is frequently used at one or more of the individual stages of a multi-stage sampling scheme – a topic which is explored below.

A SRS requires as a sampling frame a list of the sampling units – households, farmers, villages, or whatever else is being used – in any convenient order. The items listed must be numbered in sequence, starting from one for the first item at the head of the list and continuing up to as many as there are items listed. A table of random numbers or a random number generator from a computer or calculator is then used to obtain a random selection of these numbers, and the households or other items which have been given the selected numbers form the sample chosen for the survey (see Cochran, 1977:18, Caswell, 1989:7).

The table of random numbers used must have as many digits as the last number in the sample frame, and any numbers among those selected which are larger than the last number in the sample frame are simply discarded.[1] Example 3.1 gives an illustrative example of the procedure for selecting a Simple Random Sample. Like many subsequent examples, it is on a much smaller scale than any likely real instance, but is constructed on this scale in order to illustrate the procedure as simply and clearly as possible.

EXAMPLE 3.1

Consider a study of the use and effectiveness of credit by farmers who have obtained a loan from the Agricultural Bank. Suppose we have a list from the bank of all farmers who received a loan last season and can use this as the sampling frame. This list is given below, and we require a simple random sample of five farmers. First of all, the farmers listed are numbered sequentially next to their names, starting with one.

Kephas Mumba	**1**	**X**
Amos Phiri	**2**	
Joseph Mulenga	**3**	**X**
Julia Musonda	**4**	
Stephen Hamombe	**5**	
Peter Zulu	**6**	
Loveness Chipwepwe	**7**	
Jeremiah Banda	**8**	
Mutumba Mainga	**9**	
James Kazunga	**10**	
Sheila Tonga	**11**	
Benson Chimfwembe	**12**	**X**
Simon Kapwepwe	**13**	
John Njovu	**14**	
Mary Tembo	**15**	
Mpafya Musonda	**16**	**X**
Gilbert Ndhlovu	**17**	
Samuel Phiri	**18**	
Margaret Shamwana	**19**	
Robert Chungu	**20**	
Michael Mweemba	**21**	**X**

Since the last number is 21, which has two digits, we require a table of two-digit random numbers. The following numbers are taken directly from such a table:

03	**47**	**43**	**73**	**86**	**22**
97	**74**	**24**	**67**	**62**	**42**
16	**76**	**62**	**27**	**66**	**01**
12	**56**	**85**	**99**	**26**	**21**
55	**59**	**56**	**35**	**64**	**60**

Going down each column in turn, the first number is 03, which is one of the numbers in the frame. It is the number given to Joseph Mulenga, so he is selected for the sample. The next number is 97, which is out of range, being larger than the largest number in the frame, so it is discarded. The next number is 16, which is in the frame, and is the number given to Mpafya Musonda, so he is selected for the sample. The next number is 12, which gives us Benson Chimfwembe for the sample. There then follow several numbers which are out of range, and are thus discarded. The next number in range is 01, which is that given to Kephas Mumba, who is therefore selected for the sample. The following number is 21, which leads to the selection of Michael Mweemba. The sampled farmers are marked with an X in the sample frame list.

This gives the required sample size of five farmers, who are:

Kephas Mumba	**1**
Joseph Mulenga	**3**
Benson Chimfwembe	**12**
Mpafya Musonda	**16**
Michael Mweemba	**21**

The use of random numbers ensures that the sample units are chosen entirely by chance, without being influenced by any person's unconscious preferences (or even favourite numbers). In a table of random numbers, each number within the chosen range has an equal chance or probability of selection. Since therefore each of the farmers (or other elements) in the sample frame is given one number, each farmer too has an equal chance of selection for the sample.

Having used a particular type of sampling design, we must use the correct procedures and formulae appropriate to that design for estimating means, totals and other measures in which we are interested. The estimation formulae for a simple random sample are shown in Chapter 10. Details of the other designs considered in this book, and examples of calculations using them, are given in Appendix 1.

Linear systematic sample (LSS)

The procedure for SRS is very simple to apply, but sometimes it can be very time-consuming and tedious in practice. This is particularly the case where there is a very large number of elements in the sampling frame, all of which have to be individually numbered, and especially so if the items are not together on one list but are in some other form such as cards in a card index. In circumstances like these, the use of a Linear Systematic Sample (LSS) instead of a Simple Random Sample (SRS) may be appropriate, and is generally quicker to do.

An LSS requires as a sampling frame a list of all the sampling units, in the same way as for a SRS. It does not require them to be individually numbered but the count of units must be known. The total number of units in the frame is divided by the required sample size to obtain the sampling interval, say k $(=N/n)$. (If the calculation does not give a whole number, the next smallest whole number is used.) Then a table of random numbers is used to obtain one random number between 1 and k inclusive,

---

EXAMPLE 3.2

Let us return to the situation of Example 3.1 but select the sample using LSS instead of SRS. We therefore do not need the farmers in the list to be individually numbered, but we do need to know the total number of them, which is 21. We require a sample size of five, and so the sampling interval k is to be $21/5 = 4$, taking the next smallest whole number. We now require a random number between one and four, and for this we need a table of one-digit random numbers. The numbers below are taken from such a table.

The first number in the table is 5, which is too large and is therefore discarded. The next number, 6, is also discarded, but the next one 3, is within the range required, so this provides our random start, r. The sample is therefore provided by the third farmer in the list, and every fourth farmer after that. Going through the list in this way, we obtain the sample:

Table:

5	3	7	4
6	3	3	8
3	5	3	0
6	3	4	3
9	8	2	5

$r+(n-1)k$	Sample
$r = 3$	Joseph Mulenga
$r+k = 7$	Loveness Chipwepwe
$r+2k = 11$	Sheila Tonga
$r+3k = 15$	Mary Tembo
$r+4k = 19$	Margaret Shamwana

---

say r. The random number table used must have as many digits as does the number k. The sample is then made up of item r in the sampling frame, and every kth item thereafter, i.e. items $r, r + k, r + 2k, \ldots, r + (n - 1)k$ in the frame. Thus to identify the selected sample units, it is necessary simply to count down the frame to the rth unit, and then count on k units from that, then k again, and so on until the n sample units required have been obtained (see Cochran, 1977:205).

Provided that certain conditions are met, an LSS can be treated just like a SRS for purposes of analysis. The basic requirement of this is that the list used as the sampling frame must not have any intrinsic regularity or periodicity of its own. Suppose, for example, the frame to be used were a card index of contact farmers, with say, ten farmers associated with each extension agent grouped together, and within each ten the farmers ordered according to size of their farm. Then if an LSS with a sampling interval of 10 were taken, it would be biased to include all the largest farmers, or all the medium size ones, or all the smallest ones. It would be impossible for such an LSS to include some farmers of all the different levels of size. This is obviously not a satisfactory situation, as the theory of sampling is based on it being possible to select any one of all the different conceivable samples. It is therefore necessary, when proposing to use an LSS design, to check whether the frame to be used has some built-in regularity of this kind – if it has, then it cannot be used for LSS, and it may be necessary to use SRS instead, or else to rearrange the frame to do away with the periodicity.

It is, however, perfectly acceptable for the frame to be used for an LSS to be ordered or ranked overall in some way. For example, if the frame were a list of plots along an irrigation watercourse, or a list of all the villages in a region, ordered by size, so that the village with the largest population was the first on the list, followed by the next largest, and so on, finishing with the smallest village at the end of the list, then this would be entirely satisfactory for using as a frame for an LSS. It is indeed sometimes an advantage to have a frame ordered in some way, as this can have the effect of making the sample more efficient, in a similar way to that achieved by stratification, as is discussed below.

In some circumstances Linear Systematic Sampling can be done without a formal frame being available, because the sample units arrange themselves in a suitable 'single file' order. This is the case in the example already mentioned of plots along a watercourse. A list of the plots is not necessary as the enumerator can walk along the watercourse counting every kth plot. Another common example is found in an exit poll. This is the name for a sample of units leaving a location and is derived from market research studies of shoppers leaving a store. An agricultural equivalent would be people purchasing inputs from a supply depot. A sample of these can be obtained by an investigator waiting at the exit to the depot and interviewing every kth person who leaves. Other examples of an LSS with an implicit list are the selection of cattle passing through a dip tank or vehicles along a road. In these examples the LSS is a true scientific sample and can be analysed as such.

Nevertheless, two practical points arise. Firstly, how to choose the interval, k, when the total number of units is not known, and secondly ensuring that the rate of sampling is the same whether the units are passing the enumerator quickly or slowly. On the first point, an estimate of total units may be obtainable from historical records, such as cattle dipped in previous weeks, or from previous records of sales. On the second point, the chosen interval must be long enough to permit the interview to be completed without causing subsequent sampled units to have to wait for their interview. It is important that the sample must be drawn from every kth unit and not from a fixed time interval, or from the next unit to arrive after the previous interview is completed (see Casley and Kumar, 1988:89).

Multi-stage sample designs

Single stage SRS or LSS designs have a wide range of uses. But for studies of large or geographically dispersed populations it is more convenient to use a multi-stage sample design. This is a type of design where there is a first stage in which a sample of larger units is selected, then a second stage in which, from each of the selected first-stage units, a sample of smaller units (subunits) is chosen. Further stages can also be used. A typical example is a study of cultivation practices on farmers' fields. At the first stage, villages are selected. Within each village a sample of households is taken, and for each sampled household a sample of fields. Other similar examples are irrigation channels within watercourses, plants within fields, and credit recipients within cooperative groups. The shared characteristic is the ability to reach final sampling units through a hierarchy of higher-level stages.

A multi-stage design is particularly appropriate where a large-scale survey is to be conducted, and where for logistic and organisational reasons it is convenient for the sample to be grouped together in a more limited number of geographical areas, rather than being spread thinly and dispersed across the whole country. For an investigation covering a whole country or a large region there might be hundreds of thousands of farming households and to list every one of them to create a frame would be an enormous burden on resources. A scattered sample also poses operational problems for the organisation of fieldwork. With a multi-stage design the sample units can be more concentrated and the work of preparing a frame is reduced since it is only necessary to have, at each stage, frames relating to the larger units selected at the previous stage. For convenience, this discussion will centre upon two-stage designs, but these can readily be extended to further stages where necessary.

A basic principle of scientific sampling is that every sampling unit must have a known, positive probability of being selected. But these probabilities do not necessarily have to be equal for all units. Where the probabilities are equal, the sample design is known as self-weighting, and the formulae for calculating estimates are relatively straightforward. Where the sample design is not self-weighting, then the data relating to different sample units have to be 'weighted'. Each observation has to be multiplied by the appropriate factors or weights so that all the units have equal importance in the final calculations. It is, in most circumstances, preferable to use a self-weighting design, for two major reasons. One is that the calculations involved in making population estimates, aggregating data, and so on, are very much simpler. This point is still quite important even if the calculations are being done by computer, since it is still necessary for the weights required to be correctly specified, calculated, and entered into the computer together with the instructions for applying them correctly. The second reason is that it is often the case that a survey which is designed by one person, group or organisation is processed and/or analysed by another. In such circumstances, it is extremely difficult to ensure that the appropriate weighting scheme is applied correctly – or even at all – at the analysis stage. Errors in weighting are likely to result in serious and quite unpredictable biases, as is illustrated in Example 3.3. In view of these points, the following discussion will concentrate mainly on self-weighting designs, but touching on other possibilities where appropriate.

Probably the most common and typical kind of multi-stage sample design in a rural context is one where at the first stage the sampling units are villages (or other settlement type or locality), and then at the second stage, households (or farmers, persons, farms, etc.).

In this situation, where the primary sampling unit is villages and the secondary sampling unit is households within villages, then for the design

EXAMPLE 3.3

In order to illustrate the possible effects of incorrect weighting let us consider a situation where two villages have been drawn for a sample, each with the same probability of selection, and from each village five households have been selected. Village A is a large village with 500 households. Village B is much smaller, with 50 households. We are interested in the average area of land under annual cultivation by each household, and the following areas were obtained from the sample households. We will use some standard notation for a two-stage sample. Using y to represent the value of the variable being measured, we use a subscript i to indicate the ith unit (village in this example), and a second subscript j to indicate the jth subunit (household). The ten observations from our two villages are listed here with the y_{ij} notation.

Village A		Village B	
y_{11}	3.5 ha	y_{21}	4.6 ha
y_{12}	2.4 ha	y_{22}	2.9 ha
y_{13}	1.2 ha	y_{23}	3.7 ha
y_{14}	6.0 ha	y_{24}	8.4 ha
y_{15}	1.9 ha	y_{25}	6.4 ha

The mean areas per household for the two villages are:

$y_1 = 3.0$ ha $\qquad y_2 = 5.2$ ha

We want to know the mean area overall for both villages. Taking the simple unweighted mean of the two we have a value of

$$\frac{3.0 + 5.2}{2} = 4.1 \text{ ha}$$

But village A is much larger than village B, even though the number of sample households taken was the same. For each village to be given its true weighting in the population the village mean area must be multiplied by the number of households in the village, and then be divided by the sum of the households in all villages.

$$\frac{(3.0 * 500) + (5.2 * 50)}{(500 + 50)} = 3.2 \text{ ha}$$

The mean is now 3.2 hectares, significantly lower than the unweighted value. Failure to weight the results from the two villages would have led to a serious overestimate of the true average. The difference between the weighted and unweighted values arises because the probability of selection of the households differs in each village: 5/500 in A; 5/50 in B. Weighting in effect restores the selection probability in each village to a constant value for the whole sample. Further discussion of the weighted mean is given in the context of calculating crop yields in Chapter 11.

In general notation the weighted mean is given by the sum of the mean per unit multiplied by the total number of subunits.

$$\frac{\Sigma(\bar{y}_i * N_i)}{\Sigma N_i}$$

where N_i is the total number of subunits in the *i*th unit.

---

to be self-weighting it is necessary that the overall probability of selection of each household should be the same, irrespective of which village the household belongs to. This can cause some difficulty, since it is rarely the case that the villages are of the same size. There are two main approaches to dealing with this problem. The one which is most appropriate depends on the individual circumstances of the survey.

First of all, it is necessary to consider whether the circumstances of the survey require the number of sample households to be the same for each sample village, or whether they can be allowed to differ. Generally speaking, if the survey is one where an enumerator has to be located in the sample village for a prolonged period, for example in order to record agricultural information throughout a cropping season, then it is preferable for the workloads of different enumerators to be approximately equal, and so for the sample size to be the same within each village. If, on the other hand, the survey involves a single visit, conducted by a travelling enumerator or field team, then a varying sample size in different villages causes no difficulty. This second option is the simplest situation, and will be considered first.

Two-stage sample with constant sampling fraction

In this design, the first-stage units, the villages, are selected by simple random sampling (SRS), or linear systematic sampling (LSS), as described above. Then the second-stage units, the households, are selected from the sample villages, again by SRS or LSS, but with a constant sampling fraction (not a constant sample size) at the second stage. This means that the number of sample households is the same proportion of the total number of households in each of the sample villages. If, for example, the sampling fraction is to be 4%, then for a village with 50 households, a sample of two households is selected, and for a village with 500 households the sample size would be 20 households. In Example 3.3 for village A and village B, samples of 20 households and two households respectively would have been in proportion to the sizes of the villages and no weighting would have been needed.

The probability of selecting an individual household can be calculated from the joint probability of selecting the village and then selecting the household within the village.

$$\text{Probability} = \frac{\text{No. villages in sample}}{\text{No. villages in population}} \times \frac{\text{No. households in sample}}{\text{No. households in village}}$$

Let us assume that the two villages A and B were selected by SRS from a population of 100 villages. Probability of selecting a household in village A

$$\frac{2}{100} \times \frac{20}{500} = 0.0008$$

Probability of selecting a household in village B

$$\frac{2}{100} \times \frac{2}{50} = 0.0008$$

Because this design is self-weighting the estimation of means, totals, ratios and proportions is straightforward. The actual formulae for the estimates are given in Appendix 1, together with examples of the calculations involved.

Two-stage sampling with probability proportional to size

If it is necessary for the number of sample households per village at the second stage of sampling to be kept constant, then in order for the sample to be self-weighting the probabilities of selection of the first-stage units, the villages, must differ according to their size. The villages are therefore selected with probability proportional to size (PPS) sampling. In the case of village A and village B in Example 3.3, using PPS sampling would mean that village A would have a probability of selection ten times greater than village B, because the village is ten times larger, measured in terms of household population.

The procedure for PPS is to make a list of the villages, in any order, together with the total number of households in each. Then, for each village in turn make a cumulative sum of the households in the list and record the cumulative sum next to each village. For the selection of the sample a table of random numbers is required, and this must have as many digits as there are in the final total cumulative size of all the villages together. Take random numbers from this table, and for each number identify the village for which the cumulative size is greater than or equal to the random number, but for which the previous village's cumulative size is less. (If the random number is greater than the total cumulative size of all the villages, it is discarded.) This village is then selected for the sample. Example 3.4 gives an illustration of this procedure. If any village is selected twice it is usually most convenient to abandon that sample and start again from the beginning to select another sample. But it is also quite acceptable to take instead a double sample of households from a village selected twice.

EXAMPLE 3.4

Suppose we have the following list of villages together with their individual and cumulative numbers of households. We wish to select a total of 15 households from three villages.

Village	No. of households	Cumulative no. of households	Random numbers
A	224	224	
B	573	797	449
C	1140	1937	
D	253	2190	
E	720	2910	
F	654	3564	
G	310	3874	
H	270	4144	
I	379	4523	4228
J	411	4934	4931
K	217	5151	
L	399	5550	
M	281	5831	

The first step is to select three villages using PPS sampling. For the selection of the sample, we need four-digit random numbers, because the total cumulative size, 5831 has four digits. The following is a list of such random numbers drawn from a suitable random numbers table.

4228	4931	5166
0449	1218	2790
7815	9436	8874

The first random number is 4228, which lies between 4144, the cumulative size associated with village H, and 4523, the cumulative size associated with village I. Village I is therefore selected for the sample. The second random number is 0449, which lies between 224, the cumulative size for village A, and 797, that for village B. Village B is therefore selected for the sample. The third random number is 7815 which is too large and is discarded. The next is 4931, which corresponds to village J in the list, and that village is accordingly selected. For clarity, the random numbers are shown in the table above in the appropriate places corresponding to the cumulative sizes. The final sample of three villages therefore consists of villages B, I and J. A sample of five households would then be taken in each village using SRS or LSS methods.

The probability of selecting a household in each of the three villages is calculated as follows:

Probability of selecting village × Probability of selecting household

$$\text{Village B:} \quad \frac{573}{5831} \times \frac{5}{573} = 0.0009$$

$$\text{Village I:} \quad \frac{379}{5831} \times \frac{5}{379} = 0.0009$$

$$\text{Village J:} \quad \frac{411}{5831} \times \frac{5}{411} = 0.0009$$

The selection probabilities are the same in each village, and hence the design is self-weighting.

When the sample villages have been selected by PPS, then a fixed number of sample households per village can be selected from the sample villages. These can be selected by SRS or LSS, according to the procedures described above and illustrated in Examples 3.1 and 3.2.

The estimation formulae for this sample design are quite similar to those for the previously-described design with proportional sampling. They also are given in Appendix 1.

Choice of two-stage design

We have considered two types of two-stage design. The first used simple random, or linear systematic sampling at both stages, and is described in shorthand as SRS/SRS. The number of subunits at the second stage was selected in proportion to the total number of subunits in the first-stage unit. The second design used probability proportional to size sampling at the first stage, followed by SRS or LSS at the second stage. The number of subunits at the second stage was constant for all first-stage units. This design is termed PPS/SRS. Both designs are self-weighting and both are frequently used.

PPS/SRS is particularly appropriate for a continuous or multi-visit survey where enumerators are stationed in a fixed locality and it is desirable to make their workloads roughly equal. To be effective, the PPS design does require reasonably good information on the size measure being used. If data are not available, or are badly out of date, the first-stage selection probabilities would not be accurate and the design would be inefficient.

The SRS/SRS design does not require any prior information on size to be available and in general needs less time to be spent on frame preparation and sampling, particularly when linear systematic sampling is used. It can also be used in situations where the size of the primary sampling unit cannot be known because the unit is only formed during the course of the survey. An example of this is a study to investigate the items purchased by customers of an input supply depot and their knowledge of how to use them. A design would be first, to select a sample of depots by SRS or LSS from a list of depots. A sample of customers leaving the depot would then be taken using LSS for an exit poll.[2]

Sampling with and without replacement

With some of the sampling schemes we have discussed it is possible to select the same first- or second-stage unit twice. This is because sampling is done *with replacement*. When a unit has been selected it is left in the pool of units which are being sampled and therefore could be selected a second time or more. It is customary to use sampling with replacement for the first stage of a multi-stage design, especially when using probability proportional to size sampling, because in some circumstances sampling without replacement can distort the probabilities of selection. Sampling with replacement also simplifies the formulae for calculating standard errors. If a unit such as a village is selected twice it may be possible to take two samples from it, especially if it is large. In most practical situations however, it is preferable for the sample not to be concentrated in a few first-stage units so the procedure would be to abandon the sample and draw a fresh one, repeating the process until no units are selected twice.

For a single-stage sample, or the second stage of a two-stage sample, sampling *without replacement* is usual. When a unit is selected it is removed from the pool of units which are being sampled and so cannot be selected twice. In these circumstances there is no distortion of probabilities and the only effect on the formulae for calculating standard errors is the introduction of a factor known as the *finite population correction.* This is a multiplying factor which is introduced to correct for the fact that the sample is drawn from a finite population, rather than an infinite one. In practice, provided the sampling fraction is small, i.e. the sample size is small compared with the population, then the finite population correction can be left out of the calculations. The usual rule of thumb is that if the sampling fraction does not exceed 5% the correction can be ignored (Cochran 1977:18,24).

Cluster sampling

There are some types of survey where it is necessary or convenient to have contiguous groups, or clusters, of households or other second-stage units. This is the case, for example, in surveys investigating vital events such as births and deaths, or migration, where the focus of interest is on the occurrence of individual events within a particular carefully-specified locality. It is also often convenient for logistic reasons to have sample households grouped fairly closely together at the second stage, so that the costs of enumerators travelling between households are reduced as far as possible. Thus for all the types of two-stage sample design discussed above, there is generally some element of clustering at the second stage, although the sample households do not usually form a cluster in the strict sense of a fully contiguous grouping. Some examples of common clusters are villages, farms along a watercourse, farmers in a cooperative or animals in a herd.

Although cluster sampling has obvious logistic and other practical advantages, it does also have disadvantages. There is usually a tendency for units in the same cluster to be relatively more similar to each other than to units in different clusters, especially if the clusters are localities. This is likely to be the case for farming households, for example, because they will have a similar cultural background, will experience similar ecological and climatic effects, and will influence each other's practices by interaction and imitation. Hence they are likely to have similar cropping patterns, and to follow similar agronomic, farm management and economic practices.

It is possible to measure the extent of this effect, and the measure used is known as *intra-cluster correlation*. This measures the extent to which elements in the same cluster are more (or less) highly correlated than elements in different clusters. If the intra-cluster correlation coefficient for a variable is positive, this means that sampling clusters or groups of households is less efficient than simple random sampling of households spread evenly over the whole population. The larger the size of the clusters, the less efficient is the sample. For agricultural characteristics, the level of intra-cluster correlation is often quite high and it is therefore advisable to keep the cluster size, such as the number of sample households per sample village, as small as possible. If no information is available concerning the actual size of intra-cluster correlation coefficients in a particular situation, it is probably wise to aim for a maximum cluster sample size of around five, as a convenient rule of thumb. In a sample design involving static-based enumerators it may be necessary for logistic reasons to have a larger cluster size and to accept the penalty of lower efficiency, but every effort should be made to keep it as small as possible (Casley and Lury 1981: 80, Poate and Casley 1985: 29).

Stratification

Stratification is the dividing up of the survey universe into mutually exclusive subpopulations, which are then sampled independently. It is usually devised in such a way that the strata created are more homogeneous than the population as a whole, and if that is the case then the overall sampling variability is reduced by stratification. It is also sometimes done simply (or partly) for administrative or practical reasons, or to obtain separate results for particular subpopulations of interest. A common example of this latter type is stratification by administrative divisions such as provinces or regions or by ecological zones. Strictly speaking this is a division into domains of study rather than stratification, but such subdivisions are often ecologically or culturally homogeneous, and so still achieve the benefits of stratification in reducing sampling error.

Many different types of stratification variables or criteria can be used to divide up the population into strata. Geographical variables, such as the administrative divisions mentioned above, or division into different categories of urban and rural areas, or into ecological zones, are a very common type. Also common are characteristics of the individual units, for example, in the case of a farming household, whether it cultivates a particular crop, whether it owns a particular type of livestock, what area is cultivated, by what type of tenure the land is held, whether a particular type of agricultural input is used, whether the farmer has a particular type of contact with an extension agent – the possibilities are almost inexhaustible.

In principle, the actual mechanics of selecting a stratified sample are quite straightforward. The population is divided up into the various strata, and then each stratum is sampled individually, using whatever sample design has been selected – stratification is basically an optional factor which can be superimposed on any sample design. In practice, actually doing the stratification can sometimes be more difficult. If a characteristic of the individual household is to be used as the stratification variable, then the information relating to this must be gathered when the sampling frame is prepared, for all the households listed in the frame, and must be recorded as part of the frame. The form illustrated in Example 2.1 can be used for this purpose. This is likely to be a large task, and if the information gathered is at all complex it may be difficult to ensure that it is done well, especially if the field staff are at this stage newly-recruited and inexperienced. It is also possible that some types of stratification variables may change between the time of frame preparation and the survey fieldwork, so that an additional phase of frame updating may be needed.

For the types of two-stage sample designs discussed above, stratification may be done at either (or indeed both) of the stages. If it is done at the first stage, then first all the villages are assigned to the appropriate strata, and a sampling frame of villages constructed for each stratum separately. Sampling then proceeds for each stratum independently, according to the designated sample design. If there is stratification at the second stage, then first of all the first-stage sample units are selected, then for each of these a set of sampling frames is constructed, one for each stratum, and the sample households for each stratum selected separately, according to the chosen sample design. An example of a two-stage design with stratification at both stages is a sample of villages stratified by ecological zone, then of farmers stratified by the number of oxen owned.

In order to obtain an estimate of a population parameter from a stratified sample design, it is necessary first to obtain the appropriate estimates from the individual strata, and then combine them to give an overall estimate. The formulae for calculating the overall estimates and for their standard errors are given in Appendix 1. These calculations are simpler if the sample is self-weighting. This occurs if the sampling fraction is the same for all strata, i.e. if the sample is allocated between strata in the same proportions as the sizes of the strata.

When considering stratification, the question arises as to whether it is necessary, and whether it is worthwhile. Will it result in significantly more homogeneous subgroups, with a consequent reduction in sampling variability? If so, will the improvement be worth the extra time and trouble in designing the sample, preparing the frame, and doing the analysis? Furthermore, will it be practicable to do the stratification successfully, and produce good quality data? Stratification at the second stage is often not practicable, especially if the definitions of the strata are complicated. Finally, if stratification is decided upon, it is useful to ask whether the design can be made self-weighting. If some strata are much smaller than others, or much more variable, then this is not likely to be feasible, but in the absence of factors of this kind then a self-weighting design is very much preferable, not least because of the greater ease and convenience of the calculations relating to this type of design. Also, however, problems sometimes arise with stratification variables during the course of a survey, perhaps because the frame preparation has been faulty or not updated, and in such a situation a self-weighting design is more likely to be retrievable as a non-stratified design (Casley and Lury 1981).

Replicated sampling

In experimental work, it is frequently the case that samples are replicated, or repeated several times, in order to throw more light on the effects of different factors, and the variability associated with different sources. This is not usually feasible in survey work, but there are some related approaches which may be appropriate in certain circumstances.

Firstly, in some data collection techniques, an element of replication may be useful. This will be discussed further in Chapter 5, but a typical application would be in measuring or weighting a subsample of a crop harvest, such as the produce of a fixed-size subplot, or a standard local harvest unit such as a heap or bundle. There is likely to be an element of random variability in such units, and so replication is advisable.

A different application arises in the use of inter-penetrating samples. This technique was developed for studying the effects of some types of non-sampling error. If, for example, it is thought that the results of a survey may be affected by biases associated with different enumerators, then the sample selected for the survey is divided at random into subsamples to be allocated to the different enumerators. Since the subsamples are selected at random, it is possible to estimate the effects of the enumerator biases separately from the effects of other sources of error. There are obvious logistic problems associated with the use of this technique, in that each enumerator has to be allocated a more widespread subsample instead of a compact, conveniently located one. Where this approach is practicable, however, it is well worth considering and can be followed in more detail in Cochran (1977:388).

Summary

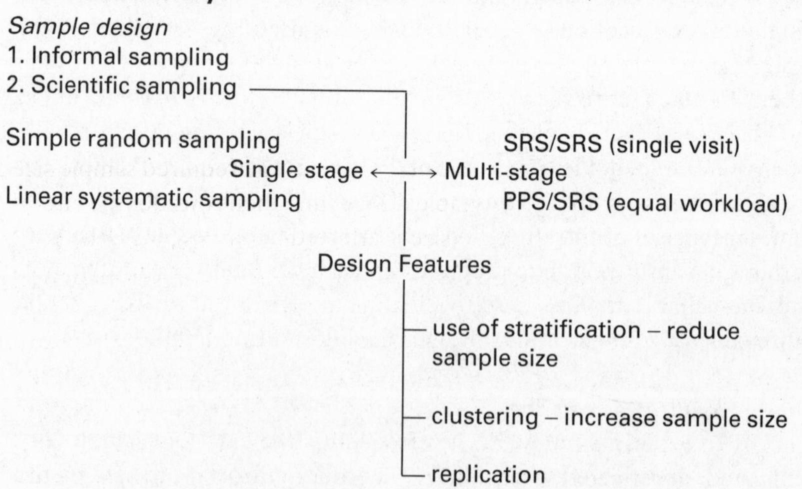

3.2 Sample size

Considerations relating to sample size

The size of the standard error of an estimate obtained from a sample survey depends upon the size of sample – the larger the sample, the smaller the standard error, and vice versa. This was demonstrated in Chapter 2. This relationship works equally in the other direction. The smaller the standard error required, and the greater the precision of the estimates, the larger must be the sample size. It must be stressed that it is the sample size, in absolute terms, and not the sampling fraction or proportion, which determines the precision of the estimates.

The sample size for a survey must be decided upon at the planning stage, together with the sample design. The sample size required depends upon three factors – the level of precision required in the estimates, the intrinsic level of variability of the variables to be estimated, and the sample design to be used. The more precise the estimates are required to be, and hence the smaller the standard error, the larger the sample size must be. Equally, if the characteristic to be estimated is itself naturally more variable in the population, then this too requires a larger sample size to attain a given level of precision. Also, as we have seen in the discussions of different types of sample design, different designs will produce different levels of precision for the same sample size, or conversely different sample sizes for the same level of precision. In general, a stratified design will tend to reduce the required sample size for a given level of precision, and a clustered design will tend to increase it.[3]

Estimation of sample size for a simple random sample

We shall consider first the estimation of the required sample size for the relatively simple SRS sample design, and then consider the effects of other types of design. Sample size is affected by three things: the level of variation in the population, the desired precision of the results, and the confidence level at which that precision is calculated.

Recall first that accuracy relates to the closeness of a sample estimate to the true population value, which we usually do not know. Precision is the closeness of a sample estimate to the mean of the sampling distribution. Luckily, this can be estimated from the properties of the normal distribution. First it is necessary to specify an acceptable level of precision. In ordinary circumstances this would be done by setting an acceptable margin of error, for example, by saying that a measurement must be made to within 1%. But an estimate from a sample is from a statistical distribution, as we saw in the previous chapter. Extreme values have some chance, even if small, of occurring. It is never possible to be 100% confident of being within a particular margin of error. It is, however, possible to specify the degree of confidence required – and the closer this is to 100% the larger the sample size has to be.

In setting the level of precision two things must be specified – the acceptable margin of error and the confidence level. A typical expression of this would be to say that the estimate should be correct to within ±10%, with 95% confidence. This means that an error greater than 10% would only occur in at most 5% of all possible samples. For example, a crop yield of the order of two tonnes, specified with ±10% error at 95% confidence would be expected to be measured within the range of 1.8 tonnes to 2.2 tonnes in 95% of all possible samples.

Probably the most common confidence level used is 95%, although 99% may be chosen when very strong evidence is required, and 90% when more moderate levels are acceptable.[4] The effect of these different levels on required sample size is shown in Table 3.1. The chosen confidence level is expressed as a value of the z statistic, from the normal distribution. Assuming that the sample size is large (greater than 30) the values of z are 1.96 at 95%, and 1.64 at 90%. For small samples ($n < 30$), the confidence level is derived from the 't' distribution and the values are larger than shown here for z.

The other factor which influences sample size is the variation in the population. This is usually denoted by the coefficient of variation, c, the standard deviation of the variable divided by its mean, which measures the intrinsic variability of the characteristic being estimated within the population. Its value is not normally known in advance, indeed it is typically one of the things which the survey is seeking to find out.

Table 3.1.

Confidence level (%)	z	Variation c (%)	Accuracy $\pm x$ (%)	n
95	1.96	50	5	384
95	1.96	50	10	96
95	1.96	75	10	216
95	1.96	75	20	54
90	1.64	50	5	269
90	1.64	50	10	67
90	1.64	75	10	151
90	1.64	75	20	38

Sometimes a rough idea can be obtained from results of surveys of similar variables in similar areas. A typical value for an agronomic variable, such as the yield of a major crop grown by peasant farmers under rainfed conditions, is in the range of 40 to 80%.

To estimate the mean of a variable, with variation $c(\%)$, confidence level z, and precision $x(\%)$, the required sample size, n, is given by

$$n = \left[\frac{zc}{x}\right]^2$$

The values of n obtained for a range of typical values of the various factors are given in Table 3.1. The table shows that in order to halve the acceptable margin of error, at constant levels of confidence and variability, sample size must be increased fourfold.

If instead of the mean of a variable, we are considering a proportion or percentage, p, then the required sample size n is given by

$$n = \frac{z^2 p(100 - p)}{x^2}$$

In this case we need some estimate of the likely value of the proportion p. If this is fairly close to 50%, then there is little variation in the values of n obtained. For a much smaller (or much larger) proportion, such as 10% or 90%, the sample size varies more widely.

The values of n obtained for a range of typical values of the various factors are given in Table 3.2.

The above calculations make no allowance for the finite population correction (f.p.c.), which becomes relevant if the sampling fraction is large. This was discussed earlier in connection with sampling with and

Table 3.2.

Confidence level %	z	Proportion p (%)	Accuracy $\pm x$ (%)	n
95	1.96	10	2	864
95	1.96	10	5	138
95	1.96	50	5	384
95	1.96	50	10	96
90	1.64	10	2	605
90	1.64	10	5	97
90	1.64	50	5	269
90	1.64	50	10	67

without replacement, where it was mentioned that ignoring the correction factor exaggerates the estimated variability. It therefore also exaggerates the sample size needed for a specified level of precision. The calculations above ignore the f.p.c., but if on occasion the calculated sample size is found to exceed 5% of the population under investigation, it may be worthwhile to allow for the f.p.c. If n is the estimated sample size before making the correction, then the corrected sample size n' is given by

$$n' = \frac{n}{1 + n/N}$$

where N is the population size.

Sample size to estimate a difference

A very common situation is where we need to assess whether there is a significant difference between the values of a variable for two different populations, or for the same population at two different points of time. Suppose we wish to be able to identify a difference of d% of the likely population values, with a confidence level defined by the value z, and the coefficient of variation of the variable of interest is thought to be about c%, then if n is the required sample size in each of the two populations it is given by

$$n = 2\left[\frac{zc}{d}\right]^2$$

When considering a possible difference between two populations, very often the direction of the expected difference is known. For example, if an improved method is being compared with a traditional one, or none at all, then we expect the improved method to produce better results. Example 3.5 illustrates this situation. In this situation the acceptable margin of error is in only one direction – it is one-sided rather than two-sided. We are concerned only if one value is greater than the other, or vice versa. This changes the values of z from two-tailed to single-tailed. In other words, the probability concerns just one side or tail of the sampling distribution rather than both. The z values are lower at the same probability for a single direction than for a margin of error in both directions. A margin of error of 5% in both directions is 2.5% at each tail

---

EXAMPLE 3.5

Suppose we are considering a project where the extension service is promoting a mosaic-resistant variety of cassava. Extension efforts are concentrated upon a subgroup of target farmers and we wish to know whether the mean yield of the new variety planted by the target farmers is larger than that planted by other farmers. If there is a clear difference then extension efforts will be expanded and intensified and the production of improved planting material will be increased. To take these steps we require good evidence that there is a difference. Evidence from research shows that yields from mosaic-resistant varieties can be 50% or more than traditional varieties, if crop management is good. We decide that a yield difference of +20% by farmers using the new variety under their own management would represent a practical level of success. We therefore wish to be able to identify a difference of +20%, and we choose a confidence level of 95%.

This is a one-sided difference to be tested, so the appropriate value of z is 1.64. We know that yields are very variable in the study area so a coefficient of variation of 70% is assumed. The sample size n for each of the two subgroups is given by

$$n = 2\left[\frac{1.64 * 70}{20}\right]^2$$
$$= 66$$

We thus require a sample of 66 target farmers and 66 other farmers.

---

of the distribution, equivalent to a z value of 1.96. A margin of error of 5% greater or less is 5% at one tail of the distribution, equivalent to a z value of 1.64. Table 3.3 sets out a range of selected probabilities and their associated z and t values. The t statistic is used when dealing with small sample sizes where $n \leq 50$. The t statistic is calculated for a particular degree of freedom (df) given by the value of $n - 1$.

Example 3.5 considers the difference between mean values of a variable from two subpopulations. In the same way it is often necessary to examine the difference in a proportion or percentage between two subpopulations. If the difference for which we wish to test is d% and the

Table 3.3. *Selected probabilities and* z *and* t *values for one-tailed and two-tailed situations*

Probability (%)			
2-tailed	1-tailed	z	t (df = 20)
68	84	1.00	
75	87.5	1.15	
80	90	1.28	1.325
85	92.5	1.44	
90	95	1.64	1.725
95	97.5	1.96	2.086
99	99.5	2.58	2.845

EXAMPLE 3.6

Returning to the situation of Example 3.5, a later study decides that the percentage of farmers who adopt the new variety of cassava is as good a measure of its success as the crop yield, and also easier to collect from a large sample. We therefore wish to assess whether a higher percentage of target farmers, compared with other farmers, adopt the new variety. The extension service, working from past experience, considers that about 10% of farmers are likely to adopt the new variety without personal extension advice and that an adoption rate of about 30% for target farmers would represent an acceptable minimum level of success from the extension programme.

A fairly stringent test is still required, because if the programme is judged to be successful it will be expanded together with increased production of planting material. Expansion involves a cost which we do not wish to incur unless we are sure adoption is really taking place. We decide to use a 95% level of confidence. We are only concerned with a difference in one direction because we are only interested to see if the proportion of adopters amongst target farmers is greater than for other farmers. For 95 per cent the value of z is 1.64.

Care must be taken over the value used for p, the overall probability. In this example we are looking for values of p from each subpopulation of 10% and 30%. Taking both subpopulations together, the percentage of adopters would be 20%, assuming an equal size of sample from both subpopulations. Sample size, n, can then be calculated from the formula.

$$n = \frac{2 * 1.64^2 * 20 * (100 - 20)}{20^2} = 22$$

Thus we require a sample of 22 target farmers and 22 other farmers.

overall proportion is $p\%$ at a confidence level z, then the sample size, n, required for each of the two populations is given by

$$n = \frac{2z^2 p(100 - p)}{d^2}$$

z values for 95% and 90% are 1.96 and 1.64 for differences in both directions, and 1.64 and 1.28 respectively for differences in one direction.

Effects of other sample designs

The effect of stratification is generally to increase precision and thus reduce the required sample size. The gains come from dividing the population into strata which are very similar with respect to the variables being investigated. If the stratification is very effective – the stratification variable covers a wide range in the population, but can be used to divide it up into convenient homogeneous strata, and the variables to be measured are closely related to the stratification variable – then considerable gains are possible, but a reduction of around 20% is a typical level.

The effect of clustering, which is likely to be one feature of the two-stage designs discussed above, is to increase the required sample size, compared with a simple random sample. For a typical level of intra-cluster correlation for agronomic variables the sample size is increased by a factor of about two for a cluster size of five. A larger cluster size than this is not recommended, since the factor becomes prohibitively large.

When estimating the required sample size for a two-stage design, it is difficult to arrive at a very precise figure, in view of all the unavoidable estimates and approximations which have to be made. It is usually adequate to make an estimate of the sample size which would be required for SRS, and then to apply an approximate correction factor to this. In most cases, the effects of intra-cluster correlation in increasing sample size are likely to be greater than the effects of any stratification in reducing it, and a correction factor in the range of 1.5 to 2 is likely to be adequate. As a convenient but rather blunt rule of thumb, when using a two-stage design, calculate the required sample size for a single-stage SRS and then double that figure.

Conflicting sample size requirements

We have concentrated on the simple situation where the demands of precision for one variable determine the sample size. But it is more usual for several variables to be of equal importance in a particular survey, and the precision requirements for each of these will then produce a different estimate of the sample size needed. In this situation, an assessment of priorities of the different data items is inescapable in order to arrive at sample size.

Another potential area of conflict is between the sample size needed to produce the required level of precision, and the resources available for conducting the survey. It may simply not be possible, with the available personnel, funds, or other resources, to conduct a survey of the size required. In these circumstances, it is necessary to decide whether to proceed with a smaller sample size and accept a lower level of precision in the results, or to postpone the survey until sufficient resources are available, which may perhaps be never.

A final point is one which has been touched on before, but which merits reiteration here. A larger sample size reduces sampling error, but it may also have the effect of increasing non-sampling error. It is highly advisable to consider very carefully the likely trade-offs between sampling and non-sampling error, and not to rush headlong into a very large, and very precise, but very biased and unreliable survey. In any consideration of survey errors, non-sampling errors should be given serious attention.

Summary

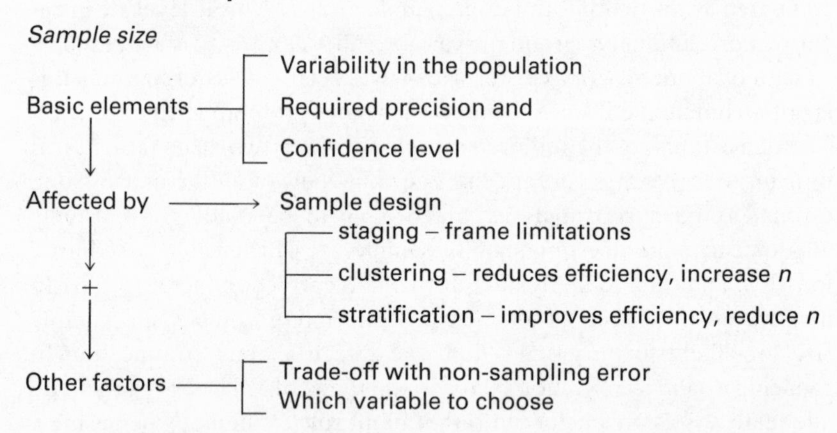

Notes and References

1. Tables of random numbers are a simple and convenient source of randomised selection for sampling. They are commonly found in the appendix to statistics texts or published in collections of mathematical tables. Snedecor and Cochran (1967:12) and Cas-

well (1989:7) include a description of their use. Random number generation programmes are also available for computers so that lists can be created when they are required. Care must be taken, however, to ensure that the numbers are drawn from the appropriate statistical distribution (uniform, implying equal chance of selection of any number in the chosen range and *not* normal).

2. Multi-stage sampling designs are more complex mathematically. Unfortunately, there tends to be a substantial gulf between the treatment of single-stage simple and linear systematic random sampling found in most basic texts, and the more mathematical treatment of multi-stage designs in more advanced texts. Useful general discussion can be found in Caswell (1989:10), Casley and Lury (1981) and Casley and Kumar (1988:92). More detailed treatment is in Snedecor and Cochran (1967:528) and Cochran (1977) who refers to multi-stage sampling as subsampling.
3. Practical guidance on the estimation of required sample size is given in Casley and Kumar (1988:83).
4. There is a tendency for surveyors to use confidence levels of 95% without real thought as to the implications of this level of chance and the use to which survey results will be put. In few instances concerned with agricultural development would expensive decisions be based on single values of test statistics. In such circumstances lower levels of confidence (with the accompanying reduction in sample size and hence cost) would suffice. A good illustration of this reasoning applied to monitoring surveys of agricultural extension appears in Murphy and Marchant (1988:31).

4

Data collection: some general issues

4.1 Unit of study

It was observed in Chapter 2 that when preparing the sampling frame for a survey it is necessary to specify very carefully the sampling units of which the frame is to be made up. This is a particular aspect of a more general issue, which applies to every type of investigation – the issue of what is to be the unit of study. As in the instance of the sampling frame, it is important that the unit of study should be precisely specified. If it is, for example, the farming household, then we must define what 'farming' means, perhaps in terms of the occupations of household members, or the area of land cultivated or owned, or the numbers and types of livestock kept. It is also necessary to define 'household' – a particularly difficult concept to deal with, as is further discussed below.

Within a survey or other types of investigation, there may well be several different units of study at different levels. Usually this will occur where the objective is to examine both a larger unit of some kind and also one or more individual subunits or components of the larger unit. Some common examples of this are households and individual persons within them, or farms and individual fields, or both multiple-cropped plots and individual crops on them. In such cases it is obviously necessary to specify each unit of study separately, and if appropriate to specify how they are related to each other.

Categories and groups of people

One of the most common categories of units of study is that involving people, either individually or in groups. A study examining individual persons could include all types of people – this might be the case for example in a study of physical disability, where the topic is of relevance across the whole population. More commonly, however, only a subgroup of the population is included in the study. Such subgroups are often defined in terms of sex and/or age. For example, a study concerned with employment might be confined to the population aged 15 and over, or a study of fertility might relate only to women aged 15 to 49.

We naturally think of grouping people in a household or family. The right type of group depends upon the purpose of the study. For example, a study involving demographic characteristics might take the nuclear family grouping as its unit of study. For many topics with an economic component, however, the household tends to be the most commonly used unit. If the subject of study involves any kind of small-scale agricultural enterprise, agricultural processing or other operations at the small family business level, then the decision-making, the organisation of labour, the economic and financial arrangements (e.g. purchase of inputs, sale of produce or use of credit), all typically operate within the context of the household. This can be defined in various ways – the group of people who 'eat from the same pot', the group of people who live together and form an economic decision-making unit, the group of people who live together and have common financial arrangements for their day to day living expenses. It usually refers to a group led by one person, the household head, who is the decision-maker for the household. It is not easy to devise an all embracing definition which will be appropriate in all conceivable circumstances, though in practice most cases will be quite clear whichever definition is used. In some societies, the situation in which, say, a group of brothers, and their families, live in a common compound, with varying levels of economic and social integration, may cause particular definitional problems. Polygamy is another cause of difficulty in defining the household. In general, it is necessary to consider carefully the prevailing social customs, and the type of functional unit which is most appropriate to the purpose of the study in question, and then try to find the most workable compromise.

Let us consider an example. In Northern Nigeria a physical compound may contain a group of people who are all related by blood or marriage. They work together for part of the time on communal land to produce food which is eaten by all members of the 'family', but also have individual or private fields which are not necessarily shared, and may take part in petty trading or other activities which again benefit the individual who undertakes the work. The compound may contain several nuclear family groups, some of which may consist of the several wives of a polygamously married man, each wife with her own children. These nuclear groups may cook and eat separately, but some or all of the food supplies may come from the communal stock. Here the economic unit is the whole compound for some purposes, such as those involving the communally-farmed land and the communal food stocks, but when considering the farming of the private fields, or the trading or other non-farm activities, then smaller units are relevant. Thus in the case of a study of agricultural activities it would be necessary to consider both the whole compound and the individual nuclear units within it as units of study, and to obtain information about the communal fields for the larger unit, and information about the private fields for the smaller units. For a study concerned with non-farm economic activities, the smaller household units would be the most appropriate.

Institutions and organisations

In some instances it may be better to take as the unit of study an institution or organisation regarded as an entity in itself, rather than as a group of people. This may be the case for instance when considering a farm, or other small business, from a strictly business standpoint, for example when studying bank credit, or national accounts. In such a case the economic or financial characteristics of the business enterprise are of interest, rather than the personal characteristics of the people owning or operating it.

Another example in this category is that of organisations which are formed of groups of people who have joined together for a specific, common purpose. Examples are a farmers' cooperative, or a water users' association on an irrigation scheme. When the subject of investigation relates to the function of such organisations, for example to the work of a crop marketing organisation servicing cooperatives, or to a scheme for the rehabilitation of watercourses, then the organisation in question is the appropriate unit of study.

Places

In surveys relating to agricultural subjects it is often most convenient to take an area of land as the unit of study. This is likely to be appropriate when investigating topics such as agronomic and farm management practices, crop production, input use or response and labour utilisation. In general, if the subject of investigation is something which relates specifically to a field or plot, or is best measured in relation to this, such as the application of fertiliser, or the use of improved varieties of seed, then the field or plot is likely to be the most suitable unit of study. If the subject is one which is more closely related to the household as a whole, for example income and expenditure patterns, or labour availability, then the household is likely to be a better unit than a land-based one.

The precise unit of study may be the individual plot of land which is cultivated as a unit, or a contiguous parcel of land cultivated by a farmer (which may be subdivided, for example into areas of different crops) or the whole area of land cultivated, operated or owned by a farmer or a farming household. The nomenclature relating to such units varies very widely and confusingly, and whatever unit and name are used, they should be defined very carefully and as far as possible conform to local custom. Information on these issues is usually best sought in an informal exploratory survey in the early stages of an investigation.

The unit of land to be used must depend on the purpose of the study. For example, if the relative prevalence of different crops and crop mixtures is being investigated, then the individual plot of cultivation is likely to be the most suitable unit. If labour use is the subject of the study, then the contiguous parcel of land may be more convenient, since it is likely to be treated as one unit by the farmer for at least some operations, such as land preparation. If crop disposal and marketing is being considered, then the complete holding of the farming household is probably the appropriate unit, if the household combines the production of a crop from all its land for disposal purposes. Most topics of investigation in this sphere require some information to be obtained from the farmer or other household members, as well as some information gathered directly from the area of land in question. It is therefore often necessary to take the farmer or household as a unit of study first of all, and then to take also one or more units of land associated with that farmer or household.

There do, though, exist types of investigation which deal directly with areas of land, or locations, as the units of study. Probably the most common example of this is a study based on linear transects. Here, either a sample of points is selected within the study area, and a line of fixed length and random bearing taken from each point, or a systematic sample of parallel lines or strips is taken across the study area. The units of study are then the transects, whether lines or strips, and measurements of the characteristic of interest are taken either continuously along them, or at a (generally systematic) sample of points along them. This technique is described in more detail in Chapter 5. It is used for investigating such topics as land use patterns, soil type distributions, cropping patterns, or distributions of nomadic people, range livestock, or wild animals. These are all topics which can be measured or assessed by direct observation. The transect method is not suitable when we also need information obtained from people by interview.

The community is a suitable unit when the subject of study relates to community-level facilities or characteristics. Some examples of these might be the numbers and availability of schools, health centres, agricultural supply depots, markets, and so on, or the provision of services such as electricity, piped water, public transport, or roads. Identifying and defining what is meant by a community poses problems similar to those in defining a household. For example, do two traditionally distinct villages, which are now contiguous and share a school and health centre, form one community or two? Does a small hamlet, occupied by an extended family of semi-nomadic cattle herders, constitute a separate community, or is it part of the nearest larger settlement, with which it is administratively grouped for local government purposes? Again, the answer depends on the purpose for which the information is to be used. The most flexible approach is to identify the smallest units individually, and then additionally to group them into larger groups. The community as a unit of study is most likely to be relevant either when considering an individual community as a case study, or when gathering gazetteer type data for listing all communities in the area of study. It is not likely to be the final-stage unit in a sample survey.

Events

A wide variety of different kinds of event may be found as units of study. Often, though by no means invariably, they are considered as subunits of study, and are observed when they occur within a higher-level unit of study, such as a household or a community.

One of the most obvious kinds is what is known as a 'vital event' – an occurrence of birth or death. Usually this is observed and studied when it occurs during the course of a more general demographic survey. Here, the household is one unit of study, but additionally each vital event occurring in the household is also a unit, to be investigated in its own right.

A very different kind of event, less easily recognisable as such but a very common unit of study, is the sale and purchase transaction. This may occur in several different contexts. Firstly, it occurs as a subunit in a household income and expenditure survey. Here the main unit of study is the household, but for the investigation of expenditure it is necessary to record and study each purchase transaction of the household. Transactions of this kind also constitute the unit of study when investigating the use of agricultural inputs by means of recording purchases from input supply depots. Price surveys are another example, where a price is collected in relation to a specific transaction (although it may sometimes be only a notional one). Here, although the collection of data takes place in the context of a trader in a market, the unit which is actually being studied is neither the market, nor the trader, but the transaction itself.

4.2 Sources of information

Secondary sources

When seeking information on any subject, the first step should always be to consider whether this information might already have been collected by someone else. Depending on the subject of study, it may be useful to consult the national statistical office, appropriate national or regional ministries, universities and research institutions, and any other bodies which may have relevant administrative records. A difficulty with using secondary sources is that any information gathered by someone else may not relate to the same concepts, definitions, and universe as are required for the study. However, such information may still provide part of the data needed, or form a basis for further data collection, or at least suggest useful hypotheses or trains of investigation.

Interviews versus measurement

There are two types of data collection – directly, by measurement or observation of the subject; or indirectly, by interviewing a respondent and obtaining his or her report of the matter. These two approaches are by no means mutually exclusive – it is very common indeed to find both being used in the same survey. Some topics can only be investigated by one or other approach, but many can be investigated using either, and in such cases it is necessary to assess which is the more suitable in the circumstances of the particular study. Figure 4.1 shows the stages which a piece of information passes through in the course of data collection. Between the information in reality, and the eventual processing and analysis, there are many transfers of data. The nature of the transfer differs between those data collected by measurement or observation, and those collected by interview. Every transfer is a potential source of error and distortion, but those involved in data collection by interview are in general likely to be the more serious, and measurement or observation is preferable, wherever possible.

First let us consider the question of whether a required piece of information can be gathered by measurement. The first question here is whether it is physically possible to measure or observe the item. If it relates to something which occurred in the past but has not left any measurable evidence behind it, then it cannot now be measured. Some examples are in seeking retrospective information on crop disposal for the last season, or on agronomic practices at an earlier stage of the present season, or on lifetime migration. Where accurate birth registration data are lacking or uncommon, the question of age also falls into this category. Another topic which cannot be directly measured is one which relates to a person's thought processes, such as questions of opinion or attitude. All these kinds of information must be approached by means of interview if they are to be gathered at all.

Information on a topic can be gathered by measurement if it is physically measurable or observable – or produces evidence which is. The area of a field, the weight of a harvested crop, the quantity of fertiliser

Figure 4.1. Data transfers.

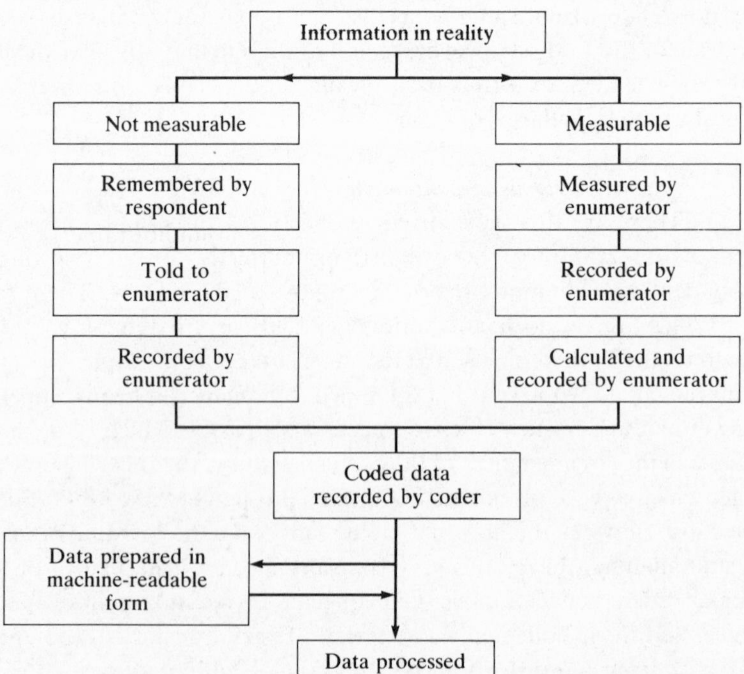

applied to a crop, and the date on which it was applied, are all items for which it is possible to gather data by measurement or observation.

Examples of topics where data are likely to be best gathered by observation and measurement, by interview, or by a combination of the two, are indicated in Table 4.1.

It is often the case that items which are in theory measurable, in practice incur heavy costs in resources to do so. For example, it is certainly possible to weigh the harvest of a crop, but it may require enumerators to be present at or around the time of the harvest, and for them to be supplied with the necessary equipment, training, and logistical back-up. If the costs of obtaining information by measurement are disproportionately large, it may be worth considering whether information on the same topic can be obtained more cheaply by interview. In the example of the weight of crop harvested, this would mean simply asking the farmer after the harvest what the weight had been, instead of weighing it directly.

Although the interview may have lower costs, it is very likely also to have a lower level of accuracy. Approximations, misunderstandings, memory lapses, or simply a lack of accurate knowledge of the topic, may all cause inaccuracies in data collected by interview. For example, suppose a farmer is asked how many hours he worked on making his yam

Table 4.1.

	Observation	Observation + interview	Interview
Population		*	
Income			*
Expenditure			*
Consumption			*
Farm labour use		*	
Off-farm labour			*
Crops grown		*	
Crop areas	*		
Planting dates		*	
Planting density	*		
Use of fertiliser		*	
Use of other chemicals		*	
Weeding		*	
Plot cultivation history			*
Crop output		*	
Crop sales			*
Crop gross income			*

heaps during the last season. He may never have known this in terms of hours, or he may give an estimate according to what is locally regarded as typical, or he may misunderstand the question and include also the time he spent making heaps for his neighbour for payment. Or he may have forgotten completely. If he wishes to get rid of the enumerator quickly, or is suspicious of him or of the survey organisation, he may make up an entirely spurious figure. None of these errors or inaccuracies would have occurred if the enumerator recorded directly the number of hours work he observed, but the cost would have been very much greater. In such cases it is necessary to consider what level of accuracy is required or acceptable, and what level of resources is available, and to match the two accordingly.

As well as the options of using measurement – more costly but more accurate – and interview – cheaper but usually less accurate – it is possible in some circumstances to use a combined, compromise approach. This involves collecting the required data by interview from the whole sample, and also by measurement from a subsample. The two sets of data can then be compared, and if necessary a correction or calibration factor can be obtained from the more accurate subsample data, to apply to the whole sample. The costs of this approach are likely to lie between those of the interview and of direct measurement, and so is the level of accuracy.

4.3 Data collection by measurement

Common types of data collected by observation and measurement include:

- land area measurement
- crop output measurement
- milk output
- animal weight gain
- instrument recordings or readings (e.g. water flow, rainfall)
- physical measurement or examination of people
- counts of human, animal and plant populations
- direct observation of work
- exchange activities (e.g. purchase and sale prices)

Data collection by measurement can be undertaken in several ways. One of the commonest is the direct measurement of a physical characteristic using an instrument. Examples are measuring the area of a field, or the weight of a crop harvested.

Another measurement activity is the observation of people engaged in an activity, and the recording of relevant aspects of their activities. This is involved in the example of measuring the number of hours worked by a farmer in making yam heaps. Some of the most common applications of this approach are in measuring the time used in agricultural operations, and in examining the distribution of a person's time between different activities. A frequent application is in measuring the household consumption of commodities, perhaps in a study of household income and expenditure, or of nutrition.

Observation and measurement may at one extreme include a variety of participant observation, where the observer or enumerator actually takes part in the activity he is measuring.[1] This is a technique most likely to be appropriate in a case study, where frequent detailed observations are of importance. An example is an investigation into the brewing and consumption of beer for a large-scale communal agricultural work party, which is a traditional means of payment. This is a very resource-expensive technique of measurement, because it requires the full-time presence of a skilled observer throughout the study. Chapters 5 and 6 consider methods of data collection in more detail and present specific techniques for measurement.

4.4 Data collection by interview

As is observed above, information may be sought by interview rather than measurement for various reasons. It may be information which could be measured directly but which would require too much time or too great a use of manpower or funds to do so. As a result the quicker, cheaper but probably less accurate interview method is used instead. It may be information about measurable events which nevertheless cannot be directly observed because they relate to the past. It may be information about matters which the enumerator cannot for social or practical reasons observe directly. Or it may be information about the respondent's own knowledge, opinions, perceptions or attitude.

There are some general problems which are common to all these types of question, and some which are more specific. In the former category are problems of obtaining the respondent's cooperation, the possibility of a misunderstanding of the question asked, possible distortion of the true answer by the respondent (for any of a variety of reasons including prestige or a wish to impress or please the enumerator), and an attitude of casualness on the part of the respondent as to whether the information given is accurate or not. On questions relating to the past, the respondent may have a lapse of memory, and have forgotten or incorrectly recalled the information. On any type of question except that relating to the respondents' own knowledge, the respondent may simply not know the answer to a question, either because it is outside his/her normal sphere of knowledge or because the question asks for quantitative information in units he/she does not use.

Some questions may cause problems because they deal with sensitive topics. What constitutes a sensitive topic varies between different cultures and societies, and sometimes between different sections of a society. Some people may find it difficult to talk about their income, or their stocks of stored grain. Others may not wish to speak of the work done by female family members. Many people are reluctant to discuss their opinions of, or attitudes to, people or institutions which are in a position of authority. Specific taboos may also cause problems, for example in forbidding people to speak the name of their eldest child, or mother-in-law, or husband. What is a sensitive topic to the respondent may not be so to the investigator, and vice versa, so it is one of the investigator's tasks to check whether any of the proposed questions may cause problems.

In order to reduce potential inaccuracies, careful consideration needs to be given to the choice of respondent, the motivation of the respondent, the circumstances of the interview and the structure and content of the interview.

Respondents

The respondent should ideally be the person who has the best, closest and most direct knowledge of the subject under investigation.

Information can be asked from:

- people as direct doers of activities
- people in positions of authority (family head, village head, politician, group chairman, etc.) about the doings of other people
- people as reporters of places or events (plots, fields, households, villages, roads, etc.) where quantities may be measured and activities take place

Wherever possible people who were directly involved in the activity should be questioned. For example, labour and income/expenditure studies of a farm family are unlikely to be successful if only the household head is interviewed. Similarly, if questions are about a place, such as a field, a visit is desirable to aid the memory of the respondent and to cross-check his/her response from visual observation.

The more 'remote' the respondent is from the subject of the survey the greater the level of inaccuracy which will arise. This can be illustrated graphically as in Figure 4.2. Accuracy is highest when information is taken from the doer of the activity. It falls when the respondent is someone in authority over the people doing the activity and is lower still if the respondent is just one of several participants when the subject of the enquiry is the overall activity.

In some types of preliminary study, where an assessment of the general situation is required, it may be appropriate to seek out for interview the local leaders, and other important and prominent persons in the community.

In a sample survey, the units of the selected random sample are the persons interviewed. If the sample units are households rather than individuals, then it is necessary to decide, and to specify to the enumerators, which types of household members are acceptable as respondents. If the required information is general knowledge within the household, for example the household composition, any adult member can be accepted as the respondent. If however some or all of the information is likely to be known with considerably more accuracy by one particular household

Figure 4.2. Accuracy and role of respondent.

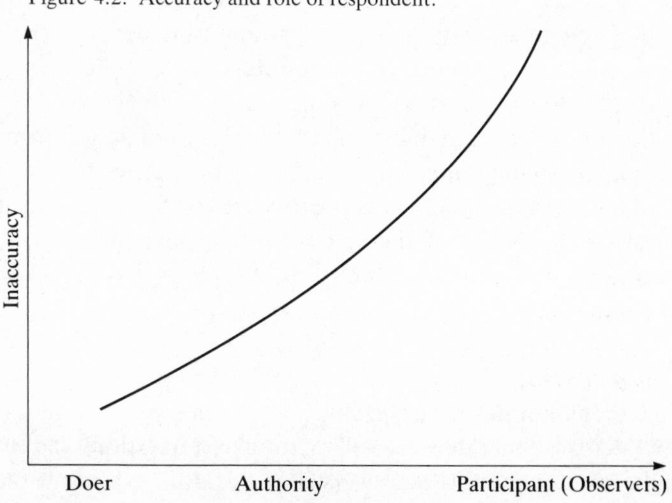

member, then it is best to specify that member as the respondent. For example, if the topic of investigation is crop marketing, then the head of household is the person most likely to have the information. There are often situations where different household members have the best knowledge of different topics within the same survey – in this situation, wherever possible, each person should be called upon to respond directly for the relevant portion of the interview. Beware of assuming that the household head has good information on all economic matters. The head may have very poor information for example on the labour contributions of other household members, even on communal land. In many societies a wife may operate a small economic enterprise within the home, for example some handicraft, or the preparation of cooked food, with respect to which only she is familiar with the details. This can happen even where the wife is in seclusion – in this case it is essential to use female enumerators to obtain good information.

In any survey there will inevitably be cases where the use of proxy respondents, that is respondents answering on behalf of someone else, has to be accepted. For example, if all the members of a sample household are temporarily absent during the whole period of field work in that locality, perhaps at a family ceremony elsewhere, then it may be acceptable to obtain some basic information on the household members from neighbours. As another instance, if a demographic survey is recording births, and a pregnant woman in the sample goes to her parents' house in order to give birth, then the information that she has in fact given birth is likely to be available from another household member. Whether a proxy respondent is acceptable depends on the type of information being gathered, and on the relationship between the proxy and the nominal respondent. This will differ with circumstances – the question to be answered is whether the proxy is likely to have good information on the topic in question. Sometimes even poor information from a proxy respondent may be acceptable in preference to none at all. In general, instructions should be given to record on the survey document the identity and position of the respondent, so that the use of proxy respondents can be identified, as well as for the purposes of checking and interpretation.

Circumstances of interview

The enumerator needs first of all to gain the active cooperation of the respondents, so that they are willing to answer questions and to think hard and carefully about the answers. For this, it is important that the enumerators are properly introduced by an appropriate person in authority, that they introduce themselves and their purpose well and relevantly to the respondents, and that they are courteous, respectful and serious in their manner and attitude.

As is further discussed in Chapter 8, the place and context in which an interview is held, and the other persons who may be present, may all affect the ease of conducting the interview, and the quality of information obtained from it. If the subject of the interview is agricultural, it is preferable for it to be conducted at the farm, so that reference can easily be made to its physical features and characteristics if necessary.

Generally speaking, interviews tend to be held among a crowd of curious family members and neighbours, and this is difficult to avoid. For some topics, for example the past cultivation history of a field, this can be useful, as the spectators may be able to assist with their own recollection. For other topics, however, an audience may be a disadvantage, since the respondent may be reluctant to admit to activities or characteristics which are low in prestige or are unconventional, or to talk at all about some private or potentially embarrassing topics. If any such topics are to be included in the interview, it is better to arrange beforehand with the respondent for the interview to be held more privately – the manner of doing so will depend on local custom.

Interview structure and content

Different kinds of interview are appropriate for different kinds of enquiries. Some general questions which must be answered when planning an interview are:

- should it be unstructured and free-ranging, or structured and limited
- should the questions be closed or open-ended (this refers to question content **not** the reference period)
- to what extent and in what circumstances can questions relating to attitude and perception be included

An open-ended flexible dialogue without formal questions is a powerful tool in the hands of a skilled interviewer and is advocated for diagnostic or exploratory surveys, for example under On Farm Adaptive Research programmes. With the right skills interviewers can probe attitudes and perceptions. But great care is needed because:

- it is very easy to lead the respondent unconsciously to answers perceived as reasonable by the interviewer

- even if not led in this way the respondent may assume that certain answers are expected and politeness results in meeting these perceived expectations
- the use of technical or obscure words or phrases may be misunderstood, creating a situation where either further explanation is needed by the respondent, or the respondent may be answering a different question from the one asked
- if, to avoid technical terms, the question is put into what is thought to be local parlance the respondent may find it patronising or, if local terms are misunderstood, misleading
- questions requiring the respondent to place himself in a hypothetical situation are difficult to answer sensibly, even by sophisticated respondents; the response will bear little relation to what would in fact be done should the hypothetical situation become real

For a large-scale survey requiring the use of enumerators an open-ended structure is not recommended. Instead a structured questionnaire with a limited range of topics will be necessary. There are advantages to using closed questions since it is easier to achieve a neat form, precoding ensures that items are not missed, and if coded data processing is to be done the coding can take place at or just after the interviews. (The subject of questionnaire design is considered in detail in Chapter 6.) Within this structured interview, each piece of information required should be well and clearly defined. Each question should have:

- a clear meaning
- a single possible meaning
- the same meaning to every person asked
- an answer which the respondent knows
- an answer which can be given clearly and unambiguously by the respondent

Reference and recall periods

Recall refers to the length of time into the past from which a respondent is asked to remember information. The reference period is the period of time during which the events being studied took place.

When a question refers to events occurring during a specified period from the past, the quality of the data obtained can often be substantially affected by the particular reference period, or recall period, to which the question relates. In general, the more recent, and shorter, a reference period, the better the information is likely to be. However, the length of the period also depends to some extent on the type and frequency of event being considered. If it is a frequent event such as the purchase of food items for household consumption, then a reference period of a few days is appropriate. But if it is less frequent, such as the purchase of durable goods, or the sale of a crop harvest, or a birth or death, then a longer period, typically in the range of three to twelve months, is usual, in order to include a large enough number of such events in the survey. There is a tendency among respondents not only to forget some events which are further in the past, but also to inadvertently shorten or lengthen the reference period for which they are giving information. It is therefore useful not only to keep the period as short and recent as possible, but also, where possible, to connect the date of the start of the period to some important and significant public event (for example Christmas, Id el Fitr, Independence Day, a general election, etc.). Some more detailed examples of reference and recall periods in relation to particular topics, including labour data, agronomic practices, and income and expenditure data, are given in Chapter 5.

Interview length and frequency

An important factor to bear in mind when planning an interview is that the burden on the respondent's time, attention and patience should be at an acceptable level. This implies that the length of an interview should not be too great. A maximum time of one hour is a convenient rule of thumb. Where a longer time is genuinely unavoidable it is advisable to make special checks on data quality and on potential response biases.

Some types of survey require repeated interviews to be conducted at intervals. This may occur for example in a survey which records agricultural activities through the course of a year. Multiple visits are needed for this in order to record the details of each phase of agricultural activity at or close to the time at which it happens. In areas where there are two cropping seasons in the course of a year, such a survey might cover either the various phases of a single cropping season, or the different seasons through the year. Another example occurs in surveys designed to collect information on the same topics at repeated interviews – income and expenditure data, or labour data, are common topics for this approach.

In a survey with repeated visits it is important that the interviews should be neither too frequent, nor too many in total. Precisely what is too frequent or too many will vary for different topics, different lengths of interview, and different types of respondent and society. Other local experience can be very helpful on this issue. It is the authors' experience that at the maximum, weekly interviews of 30 to 60 minutes **can** be conducted over a year, given **very** good field staff and a cooperative and well-motivated population – and that this pattern of interviews **cannot** be successfully conducted if either of these necessities is lacking. Three-monthly interviews, such as are appropriate for some types of demographic or labour force surveys, seem to be acceptable in most circumstances, and can be continued over three to five years.

Sample retention

In the case of a survey programme which is designed to continue over several years, it is necessary to consider whether to retain the same sample throughout, or whether and when to change it. Such a survey programme is usually concerned with estimating the values of some characteristics on a current cross-sectional basis, and also with estimating changes over time. For considering changes over time, the precision of estimation is much improved by retaining the same sample. For example, if an investigation is concerned with estimating the increase over time in the proportion of farmers using an improved variety of a crop, then it is most efficient to retain the same farmers in the sample from year to year. Keeping the same sample over several years does not adversely affect the current, cross-sectional estimates for any particular year, though it does make it impossible to group together several years' samples in order to obtain more precise cross-sectional data for characteristics which do not change over time, (an issue which is unlikely to be important in the kinds of surveys we are considering). It is therefore preferable where possible to retain the same sample, subject to the considerations of possible respondent fatigue discussed above. In most surveys it is feasible to retain sample units for two or three years. If the period of the survey is to be longer than that it is wiser to replace the sample after that period. In that case, a phased or 'rolling' programme of sample replacement is preferable, with perhaps a half or one-third of the sample units replaced each year. This ensures that for any pair of adjacent years some part of the sample is the same, giving added precision to the estimates of change between the two years (Scott 1985:42).

Notes

1. Chambers (1983) describes data collection techniques, which include social anthropology, from a critical perspective. A variant of participant observation in the context of beneficiaries of development projects is put forward by Lawrence Salmen (1987).

5

Data collection: Methods I

In Chapter 1 we emphasised the need to match the scale of enquiry, sample design and collection method to the objectives of the survey. This implies that there is scope for choice in the methods available to surveyors. For some items, such as land area and crop output, there is. Different methods can be applied according to the requirements in terms of scale and accuracy, and the enumeration resources available. For other items there may be less choice, but important considerations exist about how to collect the data. In this chapter we review specific collection methods for three important topics: crop output; crop area; livestock production. Farm inputs, population, labour, household assets, income, expenditure and consumption, and price data are considered in Chapter 6. The space alloted to each topic is not equal. Considerable attention is given to crop output and land area, because these two items are often the principal objective of surveys and are used as the dependent variable in many analyses. A wide choice of methods also exists for these items, with important bearings on scale and accuracy. Area and output are both measurement-intensive. Other topics are based around interview techniques, and draw on the principles set out in Chapter 4. Throughout the chapters the emphasis is on data collection from subsistence farmers with occasional reference to cash crops and pastoralists.

5.1 Crop output

Estimates of crop production are essential statistics for scientists and administrators concerned with research, planning and evaluation of agricultural investments. The expansion of development projects with their requirement for detailed planning has given a new prominence to crop statistics. Scientists working on farming systems research and project evaluators want to be able to produce accurate and reliable estimates of crop production from selected farmers. Each use will have its own requirements in terms of the desired accuracy and statistical precision which the results must attain. Amongst other things, these will be affected by the sampling scheme and scale of enquiry of the study. But most important of all, is the method used to collect the data.

The techniques in common use follow logically from the problem. The most straightforward approach would seem to be to ask the farmer. This may present problems, depending on the farmers' ability to express output in measurable units. Alternatively, the enquiry may need objective, standardised measures. The best objective estimate of output from a plot of land would be to harvest the complete plot and weigh the produce. If considerations of plot size and volume of output limit the number of plots which could be handled in this way then the output from the plot can be sampled, either before the farmer makes his own harvest (the crop cut) or after he has harvested (sample of harvest units). Another approach when marketing is controlled is to obtain purchase records from the marketing agency. This method is particularly important for tree crops such as coffee, cocoa, and rubber. There are therefore at least five potential techniques for estimating crop output. Each technique has its limitations. Intending crop surveyors must match the technique to the purpose for which the data are being collected. This chapter describes each technique and the circumstances in which it is best suited.

Measurement of crop output is the subject of a number of guides and manuals. FAO (1982), Poate and Casley (1985) and Poate (1988) review techniques from the perspective of national statistics, and monitoring and evaluation. We do distinguish between crop output and crop yield in this chapter on methods. Yield is defined as output per unit area. All the methods described for measuring output can be used for yield. In most practical circumstances output is what is measured in the field and yield is one aspect of the analysis of output. A number of important issues about the analysis of crop yields is presented in Chapter 10.

Farmer estimates of output

The idea of asking a peasant producer to estimate his crop output is commonly dismissed as being unscientific and prone to distortion. But on the limited evidence available, the method may be no more biased than crop cutting, which is thought by many to be the preferred objective technique.

The method depends on the existence of a consistent unit in which farmers collect their harvested output. This may be a traditional measure such as an akumada (a goat skin volume measure in parts of Ethiopia) or a modern hessian sack. But it must conform to accepted dimensions and be widely used in the farmers' community. The farmer is asked to report the number of units harvested, which can then be converted to metric measure by taking a sample of such units from the farmers' locality and calculating a conversion factor.

In using this method the surveyor must ensure that crops are reported at a consistent state of harvest, and that the conversion factor applies precisely to that state. Thus, for example, if the crop is maize all observations should be either for grain on the cob or for shelled grain, and not a mixture of the two.

The method can be used for individual estimates of plots or fields, or for estimates of the total production from the holding.[1] If a field or plot estimate is required, it is preferable for the farmer interview to take place at the field, to ensure there is no confusion over boundaries and the interview needs to be done before the output from the relevant area has been amalgamated with produce from other areas of the same crop. This places an important limitation on the approach, especially under mixed cropping and it may be preferable to restrict estimation to the level of the holding. If a holding estimate is required, the interview is better carried out at the farmer's house after inspecting his fields, when it may be possible to inspect the harvest in store. If crop yield (output per unit area) is required, then clearly an area measurement will be necessary.

Critics argue that the farmer will be inclined to distort his or her response, either to conceal output from officialdom, or to inflate performance to qualify for credit or other favourable treatment. Concealment applies equally to more objective measures which ask the farmer to identify all his fields to the surveyor. Panse, writing in 1958, reports the phenomenon of growth in reported holding size by length of residence of the enumerator. For a surveyor in a hurry the only solution is careful briefing and cooperation from the farming community. A second criticism is that the farmer will not know his output sufficiently accurately. Considering that for many people harvested output is the key to survival for another year, the idea that its quantity will not be known is hard to accept.

The merit in the technique depends on the level of bias between the farmers' reported figure and the true production. A study in Nigeria (reported under crop cutting, below) included a farmer estimate of output which indicated a bias of +14% in bullrush millet. Work on rice in Bangladesh confirms the same order of magnitude (Greeley 1987), and studies using this method have produced mean yields of the same order of magnitude as other objective methods. A joint World Bank, FAO, and UNICEF study (Verma *et al*. 1988) produced estimates between –8% and +7% in five African countries. A level of bias of less than +15% would rank this method alongside crop cutting, but with the advantage of greater simplicity and needing fewer resources for a given scale of enquiry.

If traditional harvest techniques permit this method to be used it should be considered as a valid and cost-effective approach for estimates of local or regional production, or of total output per farm family. In this context, in view of its low cost and simplicity, it is a serious alternative to crop cutting.

Whole plot harvest

A proposal to harvest the whole of a farmer's plot may appear at first to be impractical. The physical output from a small plot is significant and measurement from a complete holding would be unreasonable for more than a very small sample. But the potential benefits are considerable. The crop sampling methods described below have disadvantages. The bias associated with crop cutting severely circumscribes its use for input–output studies of crop response, and under situations where within-plot variation is as high as between-plot, the crop cut may produce a very poor estimate of plot output. Sampling of harvest units is also not reliable for plot estimates. If high quality results for individual plots are necessary for the purpose of the study, whole plot harvesting should be considered for small-scale investigations.

In common with all the objective measures, the method requires the enumerator to be present at the time the farmer wishes to harvest his plot. Unlike crop cutting of subplots however, farmers have been reported as finding this method unintrusive. They have to harvest their plot at some time and all that is asked of them is that they cooperate with the enumerator to weigh the crop. At the worst, if the enumerator is not present, they might have to delay harvest for a day or two. In many surveys such cooperation is given willingly. There are no specially demarcated areas to get in the farmer's way and disrupt his work, and any assistance by the enumerator will be a help to the farm labour force. The output measured is from an area with which the farmer is acquainted and in some instances farmers take an active interest in the results obtained.

Most important of all, there is no bias in the results, although the data may reflect neither biological yield nor economic yield, unless losses in transport to the farm and in storage are taken into account. Unless the enumerator takes control of the harvest away from the farmer, field losses will be the same as under normal farming practice.

Sampling harvest units

This method samples output after the farmer undertakes a normal harvest. It relies on the farmer collecting the crop together in some sort of a unit: bundles, sacks, bowls, or such like. But unlike the farmer estimate described above, it does not matter if the units are peculiar to each individual farmer. The enumerator visits the farmer at the harvested plot and inspects the units. Unless the number harvested is small and can all be weighed, a sample of units is taken and weighed. Total plot output is estimated by calculating the mean harvest unit weight and multiplying by the total number of units harvested. Crop yield for the plot is calculated by dividing by plot area. The technique is straightforward and can be used for large sample surveys, but a number of practical problems limit its usefulness.

The technique depends on knowing accurately the total number of units harvested, and drawing an unbiased sample from those units. Harvest is a busy time, and for many crops it can extend over a number of days. Unless the crop is left in temporary storage at the plot it may be removed to the farmer's house at the end of each day. If this happens the enumerator cannot be sure either of inspecting the total output from the plot, or that the units have not been mixed with harvest from another plot, unless he attends each harvest personally. Many crops are not harvested in one operation, but gradually over weeks or months: cotton, vegetables, and roots such as yam and cassava follow this pattern. In this situation, the possibility exists that the enumerator will try to compromise on the problem of drawing a sample by weighing just one or two harvest units, and imputting the weights of the rest of the required sample.

Mixed cropping presents further difficulties for the identification of plots. Consider a situation which is common in the Guinea Savanna zone of West Africa. A field is planted to short-season millet at the start of the rains. Later, sorghum is planted over half the field, and groundnuts on a further third. To the surveyor, each separate mixture constitutes a different plot for which separate output estimates are required. But the farmer will harvest each *crop* as a single entity without concern for the surveyor's plots. Unless the enumerator is actually present to divide the crop between each plot mixture it is unlikely that accurate plot estimates will be made. For successful plot estimates of yield from this method, three separate measurements must coincide correctly: the number of units harvested; mean unit weight; and plot area. The authors' experience with sampled harvest units on large-scale surveys in Nigeria suggest that plot estimates are unreliable but that estimates for the farm holding (which smooth the inconsistencies between plot records) are usable.

The technique is suitable for use on large-sample surveys, but is preferable where estimates for farm holdings, rather than plots, are required. It is not recommended for crops with multiple harvests or root crops harvested as individual tubers.

Crop cutting

Crop cutting involves the measurement of output from one or more yield subplots (YSP) laid in the farmers' plots under study. Crop yield can be computed directly without area measurement, because the area of the cut is known. The technique was pioneered in India and is in widespread use, being commonly regarded as the most reliable objective method. But despite its popularity, many doubts have been expressed about the accuracy of results. Modern texts such as Casley and Lury (1981) caution against the use of crop cutting, especially for large-scale studies. In Africa, Verma *et al.* (1988) found overestimates of around 30% and conclude that farmer estimates can be superior. The main issue is the level of bias which can ordinarily be expected and the demands on supervision to achieve acceptable accuracy in sample surveys.

The most frequently quoted sources of bias are due to edge effects (inclusion of plants within the YSP which actually fall outside it); border effects (a tendency for the border of the plot to have a lower chance of inclusion in the YSP because of the rules governing location); and location effects (a tendency for the enumerator to bias the location of the YSP away from low-yielding areas in the plot). The technique overestimates yields in most instances. A number of factors contribute to this. Because the YSP is normally harvested by the enumerator who works intensively over a small area, unlike the farmer who has to deal with the whole farm, the measurement will be closer to biological yield rather than economic yield, even though the needs of the survey will more often be economic yield. Allowance for losses must be made. Other issues relate to the size of the YSP, the number of YSP laid in each plot, the shape of the YSP and the rules for locating it. These are reviewed in turn.

Size of the YSP

The issue of size implies a trade-off between level of accuracy and the demands of harvest. Pioneering work by P.C. Mahalanobis on jute and paddy rice, and by P.V. Sukhatme on rice and wheat established a body of evidence that a moderate size of subplot would produce yields with a low level of bias. The most commonly quoted results are reproduced in Table 5.1. Small plots over-estimate yield but the degree of over-estimation becomes smaller as plot size is increased. From the table it could be inferred that for all practical purposes, a YSP of 40 m^2 or greater would produce results with no effective bias.

The experiments reported in Table 5.1 were carried out on plots at research stations, and the crops were grown in dense, evenly planted stands. Farmers' plots are more often uneven in planting density. FAO (1982) reviews the evidence and concludes that the size of the crop cut should be a function of the density of crop within the field, arguing that the variability of yield per plant within the same field is generally low. The paper suggests that for very dense crops 'the plot size could be quite small 1–5 m^2. For more widely spaced crops like maize, tubers, etc., the plot size could be larger 10–25 m^2. While, for very widely spaced crops and in

Table 5.1. *Overestimation of yield with small plots*

Type of crop and shape of subplot	Size of subplot (square metres)	Number of plots	Overestimate (%)
Wheat-irrigated			
Equilateral triangle	43.80	78	0.0
Equilateral triangle	10.95	78	4.8
Equilateral triangle	2.74	78	15.7
Circle	2.63	117	14.9
Circle	1.17	117	42.4
Wheat-unirrigated			
Equilateral triangle	43.80	107	0.0
Equilateral triangle	10.95	107	11.0
Equilateral triangle	2.74	107	23.4
Circle	2.63	162	14.8
Circle	1.17	161	42.4
Paddy rice			
Rectangle	40.47	108	0.8
Circle	2.63	216	4.5
Circle	1.17	216	9.0

Source: Sukhatme, P.V. (1954).

the case of mixed cropping, the plot size could be as large as 100 m^2.' (para 344). The likely bias at these sizes is not discussed.

In the absence of firm guidelines a wide variety of sizes has been used. In Nigeria, a nationwide evaluation survey programme in 1980 adopted a triangular shaped cut of 100 m^2 for all crops. But concern about the suitability of this size led to a study of YSP bias in yam and sorghum. The study involved harvests from combinations of one or more 50 m^2 triangles and 100 m^2 squares together with a total harvest from the whole plot. Farmers' fields were used for the study. Overestimates as high as 28% for sorghum and 17% for yam were found when only one 50 m^2 triangle per plot was used. There was little improvement until the area sampled increased to 200 m^2. (Molokwu and Poate, 1981)

Error from this or any other sample method is a combination of sampling error and the bias inherent in the technique. Sampling error can be reduced by increasing the number of YSP harvested in each plot, so the key issue is the bias. The Nigerian data were later reanalyzed in Poate and Casley (1985) using a variant of the standardisation method. The resulting analysis showed mean errors of the order of 8–10%, with no significant differences between the 50 m^2 and 100 m^2 subplot sizes. Supporting evidence for this order of magnitude has been reported by Greeley (1987) from work with rice in Bangladesh.

This conclusion is important. The Nigerian study was carried out under careful supervision, more so than could be expected under practical survey conditions. In these circumstances the bias is higher than would be hoped for. But if the average bias is consistent in magnitude and direction, the technique would still be usable, subject to correction during analysis.

Weighing the output

The size of subplot has a bearing on the harvest weight to be measured and therefore the scale to be used. A plot measuring 50 m^2 in a crop with an average yield of 700 kg/ha, would yield 3.5 kg. In an area with a wide range of crops such as cowpea and cassava, ranging in yield from 100 kg/ha to 10 tonnes/ha, the sample harvest weight will vary enormously. Accuracy is the key objective and the scales should be chosen for both scale and ease of reading. They must also be robust enough to survive life in an enumerator's field bag. A suspended spring balance with a large clock dial, capable of weighing up to 25 kg in units of 100 grams is a good choice for all except heavy root crops. It is important that the scale has provision for adjustment of the pointer to the zero point by a thumbscrew. Crops have to be weighed in a container, usually a sack, and it is convenient to calibrate the scale to zero with the empty sack, so that crop weight can be read directly from the scale. If the scale cannot be hung from a tree, or the rafters of a building, or a vehicle, an alternative is to suspend it from a pole held by two people.

Number of subplots

The issue of average bias is relevant when the results are to be applied to regional or crop specific averages. But in many instances the surveyor will be concerned not just with overall production, but with the output from the plot in relation to measured inputs on that same plot. In this situation the crop cut must give an unbiased estimate of the parent plot.

If the crops are planted in dense, evenly spaced stands, variation within plots would be expected to be lower than variation between plots. Data from the Nigerian crop cutting study were analysed together with independent data from Niger, and the results are shown in Table 5.2. The variation within plots is at least 40% for the three crops shown, and rises to nearly 60% in the case of yam.

In view of the level of within-plot variation it would seem that more than one YSP must be laid if reasonably accurate estimates of whole-plot yield are needed. The sampling error decreases in proportion to the square root of the number of subplots laid. Thus the sampling error from two subplots will be 70% of that from one; and the error from three will be 58% of the error from one. Given that there would not seem to be any reason to increase the size of the subplot above 50 m^2 a practical rule

Table 5.2. *Variation between and within plots*

Country and crop	Subplot size (m^2)	Number of subplots per plot	Number of plots	Variation between plots (%)	Variation within plots (%)	Total variation (%)
Nigeria						
Sorghum	50	6	30	55	45	100
Yam	50	6	31	42	58	100
Niger						
Millet						
1982	30	3	99	60	40	100
1983	30	3	103	52	48	100

Source: quoted in Poate and Casley (1985).

would be to lay two or three YSP. However, there is more work involved in laying multiple YSP than in laying larger YSP, due to the need to locate and demarcate each subplot. Therefore it would be tempting to argue that the easiest solution would be to lay two or three contiguous YSP, in effect making one larger subplot. Unfortunately, the evidence about the within-plot variation supports the need for separated YSP.

Analysis of the Nigerian subplot data by Poate and Casley compared deviations of subplot yields about their means for sets of two separated and two contiguous YSP samples in sorghum. The analysis reported '. . . the mean deviation of the separated subplots was 240 kilograms; the mean deviation of the contiguous subplots was 149 kilograms. Since deviations between separated subplots are some 60 percent greater than those between contiguous subplots, independently located subplots are required to improve the within-plot estimate of crop output' (Poate and Casley, 1985: 17).

The arguments about size and number of subplots lead to two conclusions. First, that 50 m^2 is a practical upper limit to the size of YSP. Second, that if plot specific estimates are required at least two independent YSP should be laid. The precise combination can be varied to suit the crops under study. Poate and Casley quote two alternatives of two 50 m^2 or three 30 m^2 YSP. The need for plot specific estimates occurs when the survey is designed to investigate the relationship between a plot input and crop yield. The only exception is where the purpose of the study is overall yield for a household or location, and not plot specific yield. In such a case the correct course of action is to lay one YSP per plot and release resources to measure more plots than would be possible if more than one YSP per plot was used.

The argument for laying more than one YSP depends on the subplots being statistically independent. Evidence from surveys where this technique has been used has shown a tendency for results from a second YSP to be suspiciously similar to the other YSP. Where multiple YSP are used the analyst should apply two tests to the data. First, is the coefficient of variation below 40% of the mean yield? Crop cut samples commonly display coefficients of variation of 50% or more. A low value indicates the possibility of fictitious data. Second, is the correlation coefficient between subplots greater than 0.7? Statistically independent subplots would be expected to have zero correlation, but in practice some correlation will be present because yield levels vary and the subplots will be correlated with their whole plots. But since variation within plots is of the order of 50% of total variation, a correlation coefficient greater than

0.7, equal to a coefficient of determination of 0.5, would not be expected from independent YSP. Errors are likely to be enumerator-dependent and the tests should be conducted separately for each enumerator's data. The tests do not prove that the data are false, they are merely a signal that further enquiries should be made.

Subplot shape

Bias arising from the shape of the YSP is thought to be due to distortion of the subplot boundaries and confusion as to which plants fall inside or outside the sample area. To minimise this, the desired shape would be one which has the smallest perimeter for a given area: a circle. Practical problems of manipulating shapes on a field, especially if the YSP is laid in a tall, growing crop, make the circle difficult to handle. The best alternative is a square. A straightforward technique is to use a rope, marked with the lengths of two sides and the diagonal of the square. The square is laid as two contiguous triangles by manipulating the rope. After pegging the corners, the sides and both diagonals should be measured with a tape to check for distortion. This method works well in mature crops although with maize, sorghum and millet it is best done when the plants are immature. At least two people need to be present to handle the rope.

If the YSP is laid in row crops, one diagonal of the square should be laid along the direction of rows. This will minimise uncertainty about which plants lie inside or outside the subplot because none of the sides will lie parallel to a row of plants.

Locating the subplot

To ensure that the yield estimates are unbiased and independent the subplots must be laid at random within the plot. The preferred method is as follows. Two random numbers are selected from tables, such that they lie in the range between zero and half the perimeter distance around the plot. (All measurements can be made in walking paces.) The first number prescribes the distance around the perimeter from a datum point A. The second is the distance into the plot from the entry point, given by the first number, to the YSP location. The field is entered at right angles to the perimeter at the entry point. If the second number is greater than the width of the plot the enumerator reverses his or her path until the YSP location point is reached. At the location point a rule must be used for laying the YSP. If the location point is close to the plot boundary, part of the YSP may fall outside the plot. The rule must cater for this and minimise the effect of giving the plot boundary a different probability of selection from the rest of the plot. The minimum distortion occurs when the location point indicates the centre of the YSP. If the YSP falls partly outside the plot, the enumerator slides it forwards, backwards or sideways, until it is fully inside the plot.

An alternative rule for location is to reject pairs of random numbers which result in the YSP cutting the plot boundary. A fresh selection is made, followed by a new location. This reduces the border bias still further, but it involves extra work and many enumerators would probably modify their pacing to avoid the need for a new selection.

Whichever rule is chosen, it is advisable that the enumerator records the random numbers used, and the location of point A on the survey form. In this way a supervisor can quickly cross-check the location during visits to the plot.

Application

To summarise, the crop cut will give results that are biased, even under conditions of close supervision; on average the bias will probably be around +10% but possibly much higher. The yield estimate is more nearly a biological yield than an economic yield and needs to be corrected for field, transport and storage losses. When plot estimates are required, two or more subplots must be used in each plot.

Crop cutting is a demanding technique for which the best results will only be obtained under close supervision. This makes it inadvisable for large sample surveys. The most appropriate use would seem to be for estimates of yields from specific cropping patterns, when plot data must be used. If overall regional or holding level production is sought, the farmer estimate is less complex and cheaper, and for the same resources would permit a larger sample size. If detailed measurement of input–output relationships is the purpose of the study, a whole plot harvest is to be preferred.

Marketing records

A number of tree crops, and sometimes cash crops such as cotton, present special problems for estimation of output. Harvesting from cocoa, coffee and rubber, for example, is spread over an extended season, with occasional flushes of higher output. Cotton is sometimes picked in frequent small harvests, and sometimes in two or three concerted efforts. When harvesting is concentrated over a short period of time the methods described above can be used, although the sampling approaches are expensive if they involve repeated, or lengthy, visits by enumerators. With some crops, such as cocoa, tree output can be estimated from a count of growing pods, but detailed inspection is necessary for this, and on-farm or processing losses would have to be estimated separately. In circumstances where marketing is done through a limited number of outlets, a simpler approach to estimating yield is to record total sales to the agency and relate them to the farmer's crop area or number of trees. This approach is suitable for economic output, but would not be suitable for estimates prior to processing.

Threshing and moisture content

All harvest estimates must be standardised to a common base. If a farmer estimate is being used, the onus is on the enumerator to ensure that all respondents quote the same state of crop. Individual allowances cannot be made for threshing or moisture content. But for the methods which involve direct harvest or a sample of harvest units, individual correction factors should be collected for each farmer or plot.

The precise techniques will vary according to the procedures normally followed by the farmers. If threshing is done immediately after harvest then measurements of total output, crop cuts or harvest units should be made from threshed produce. If threshing takes place after an extended period of standing storage in the field it may be necessary for the enumerator prematurely to thresh a sample at harvest himself, using the farmers' technique. If possible this should be avoided because threshing percentage will vary with the moisture content of the crop at threshing.

Moisture content will vary with every plot, and often within the plot. A sample of grain should be taken from each crop cut or, in the case of the whole plot, from several locations. If resources permit, the sample should be dried to a predetermined moisture content and a correction factor entered on the plot form to standardise the harvest weight. If moisture content cannot be measured it may be possible to sun-dry the sample of grain to constant weight. This would typically take up to a week.

Summary

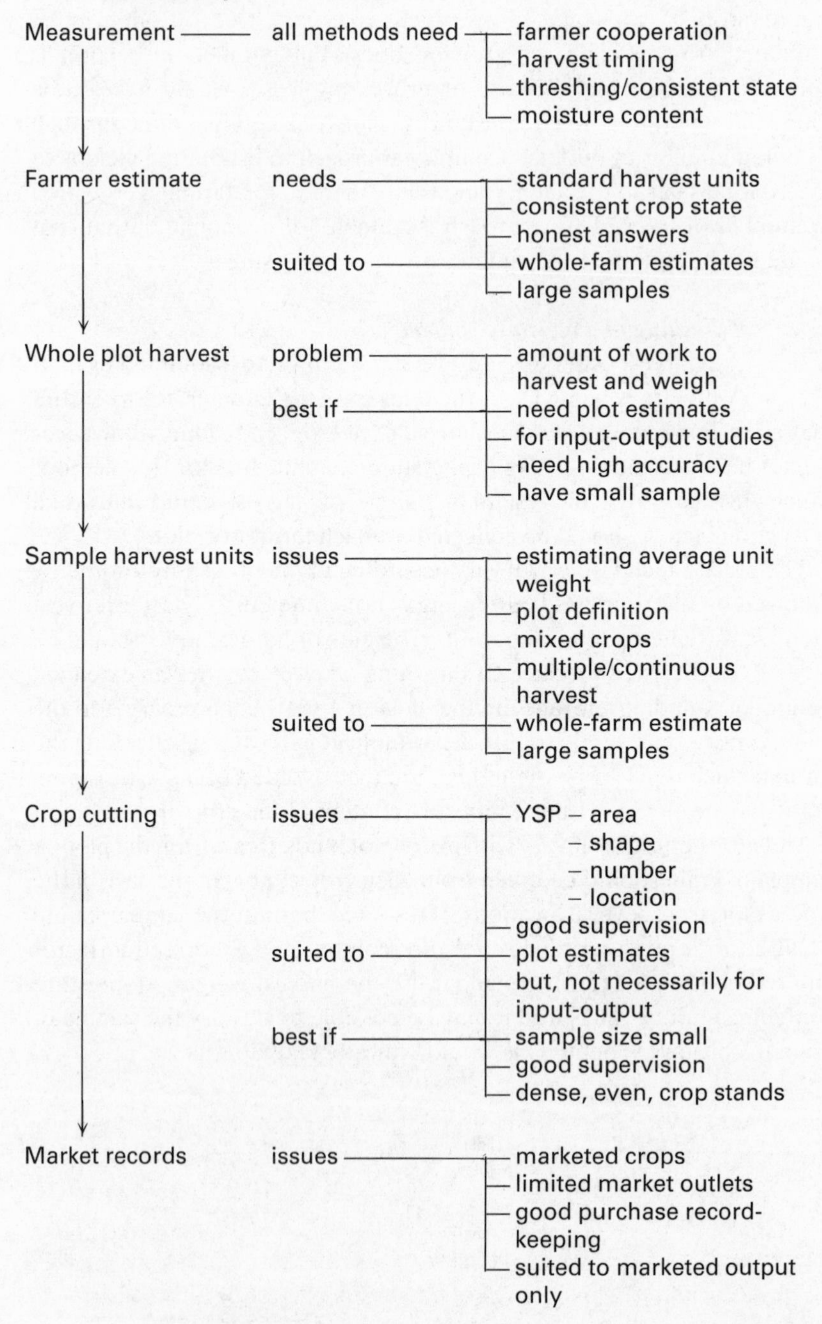
Measurement
all methods need
farmer cooperation
harvest timing
threshing/consistent state
moisture content
Farmer estimate
needs
standard harvest units
consistent crop state
honest answers
suited to
whole-farm estimates
large samples
Whole plot harvest
problem
amount of work to harvest and weigh
best if
need plot estimates
for input-output studies
need high accuracy
have small sample
Sample harvest units
issues
estimating average unit weight
plot definition
mixed crops
multiple/continuous harvest
suited to
whole-farm estimate
large samples
Crop cutting
issues
YSP – area
– shape
– number
– location
good supervision
suited to
plot estimates
but, not necessarily for input-output
best if
sample size small
good supervision
dense, even, crop stands
Market records
issues
marketed crops
limited market outlets
good purchase record-keeping
suited to marketed output only

5.2 Crop area

Statistics of land use and crop areas are amongst the most frequently collected items in crop surveys, and most countries have national agricultural survey programmes. The techniques are better known than for crop output and offer fewer real alternatives to the survey designer. There are two basic distinctions in measurement, between estimates of a region or locality, not related to individual farms, and estimation at the farm level. Regional statistics permit the use of large-scale techniques including ground transects and remote sensing. Farm-level estimates require field measurement.

Definitions

The subject of area measurement highlights the importance of clear definitions in surveys. Land utilisation is often taken as a yardstick for comparisons between localities or over time. Yet there is a danger that because of imprecise definitions comparisons are not valid. A useful review of area categories appears in FAO (1982). The broad categories of land utilisation recommended by FAO are:

1. Arable land
2. Land under permanent crops
3. Land under permanent meadows and pastures
4. Wood and forest land
5. All other land

In this structure, arable land refers to crops under some form of rotation, although in some countries permanent crops might be included in the same category. Arable land would therefore be subdivided into:

1. Land under temporary crops
2. Land under temporary meadows
3. Land temporarily fallow
4. All other arable land

But even these categories require careful description. How long, for example, is temporary fallow? How would a surveyor in central Africa treat an area of land planted with cassava as a 'famine reserve', left to fallow, and on which scrub is regenerating? Is it a temporary crop, permanent crop, temporary fallow or wood and forest? Definitions need to be stated in terms of local crop circumstances.

When data are being collected from the farm household a workable terminology is needed to describe the land farmed by the household. There are two aspects to this. One is the system of land tenure; are the fields owned, leased, sharecropped, or held under customary usufruct? The other, is the description of the land. Two classifications are in common use, both starting with the holding, a definition which embraces the farm as a whole, and includes livestock and other non-land-related farming activities.

Classification:	1	2
	Holding	Holding
	Parcel	Field
	Field	Plot

The second list has been widely used in farm surveys in Africa, whereas the concept of a parcel is more common in Asia. In both cases there is a need to differentiate between areas of land and the way in which parts of each area are managed. For example, the following definitions were used for a survey in Nigeria:

Field: A contiguous piece of land which the farmer considers to be a single entity

Plot: A subdivision of a field containing a single crop or homogeneous mixture of crops.

Plot is defined to meet the needs of the surveyor, and the boundaries may not be recognised by, or meaningful to, the farmer.

With a workable definition of the smallest unit of the farm, a plot, land farmed by the household can be measured, and the results used to assess crop areas and crop yields. But as the FAO manual points out, assessment of productivity within a plot will depend on harvested area, which may be affected by crop damage or abandonment. FAO suggest the following additional concepts in describing crop areas:

1. Area intended for planting
2. Area tilled
3. Area planted or sown
4. Area damaged
5. Area abandoned

6. Area treated with fertiliser
7. Area treated with biocides
8. Area harvested

The need for definitions is discussed in Chapter 4, and is particularly important for crop areas. The survey manual should include a section detailing area definitions and concepts, and enumerator training will benefit from practical exercises in plot and crop definitions. The example of sequential planting and a complex field of mixed cropping in Chapter 10, illustrates the practical implications of area statistics.

Regional statistics

The reason this is distinguished from household-level data (which with random sampling can also be used to generate regional estimates) is the possibility of making use of other sources, especially satellite imagery and aerial photography. Remote sensing can be used to identify different categories of land use, and in certain circumstances to estimate crop areas. An illustration of the use of aerial photography as part of a planning study was given in section 1.2. It is discussed in more detail in a separate section at the end of this chapter.

Ground transects

In situations where an overall estimate of crop distribution or land use, independent of farm holdings, is required, the ground transect is a valid method. It can be done by a small mobile team, it is fast, and above all, it does not require a sampling frame of households.

The boundaries of the area to be surveyed are defined on a map or aerial photographs. The land is then divided into a set of grid squares. Within each square a point is chosen by random coordinates. This point indicates the start position. The point is located on the ground and the enumerator walks, rides or drives a predetermined distance along a transect. The direction is chosen at random. Along the transect, observations of land use are made at predetermined intervals. The dimensions of the grids and transect are chosen to suit the enumerator resources, land area and time available for the study. In 1977 a team of 15 staff, including four supervisors, covered the 68 000 square kilometres of Kaduna State, Nigeria, in seven weeks at a cost of less than $20 000. The survey used a 10 km grid square. The transect was one kilometre in length and observations were taken every 20 metres, making 50 per kilometre. At every point along the transect the land use was recorded, including cropping details where relevant.

Land use is estimated by raising the number of observations falling on each land type to a grid total area. Thus if one of the fifty observations was a water feature, one fiftieth of the grid square area would be classed as water. An estimate of variance can be obtained by making a second transect in each square. This is often done by returning to the start point along a parallel, but separate, track. Studies in Nigeria indicated that a separation of 0.5 km was satisfactory. The formulae for estimation are simple to use (see Yates, 1949). If the transect lines are all of equal length and two lines are taken at random in each square:

$$\text{Total of a characteristic} = \left[\frac{1}{2}\sum_i (p_{i1} + p_{i2})\right] * \text{survey area}$$

where

$$\text{Standard error of a total} = \sqrt{\left(\frac{1}{2n}\sum_i (p_{i1} + p_{i2})^2\right)}$$

where: n is number of squares; p_{i1} is the proportion of a characteristic from the first transect in the ith square; p_{i2} is the proportion from the second transect in the ith square.

Ground transects are fast and cheap, but not entirely easy:

- field staff must be capable of locating start points on maps or photographs
- heavily wooded, wet or mountainous terrain can physically impede access
- difficulties with map-reading or access can result in a bias towards roads and tracks and the surrounding land use
- if the aim is to record crops grown the timing is critical to avoid missing short season, or early or late planted crops; the Kaduna study underestimated some crop mixtures compared with household surveys, due to the timing of observations
- in areas with low cultivation densities a large area of land must be covered for relatively few crop observations; stratification could be used to select crop areas, but transects are more likely to be chosen for areas where there is insufficient information about crop areas to use as a basis for stratification.

Despite these disadvantages the transect is a useful technique. It can be adapted to a variety of purposes: to sample areas on maps, aerial photographs or satellite images; to sample bounded, mapable sites such as irrigation schemes or forests; and in conjunction with low flying aircraft to cover large areas of land quickly. Its use from the air is described in section 5.4. It collects data at a single point in time. For events which evolve over a long period, such as crop mixtures in some environments, its use would be limited compared with household surveys, but for data items that occur only for short periods of time it is an attractive option.

Farm household statistics

Most surveys designed to measure crop areas are based on the farm household, landowner or operator as the unit of study. Three techniques are in widespread use and will be described here:

- use of traditional units and proxy estimates
- tape measurement
- tape and compass measurement

Traditional units are related to cultivation and the growing of crops. But the other two techniques can be used to measure other land uses such as forest or grazing.

Traditional units and proxy estimates

In many farming communities traditional measures of land area are used. The most common basis of these is the area which can be cultivated in a given time, most usually one day, either by draught animals or by hand. Where such systems exist it is feasible to ask farmers the area of plots or fields in traditional units and convert those units to metric values. For the system to operate effectively the local unit should be in use throughout the community being studied, and there should be broad agreement between members of the community about its size. It is best if enumerators using this technique conduct their interview with farmers at each plot. They will quickly gain a 'feel' for plot areas and can cross-question the farmer if an estimate appears unreasonable.

An important part of the technique is the measurement of the conversion factor. The effort put into this will depend on the purpose for which the data are collected. If the study is part of a rapid reconnaissance, or is concerned with regional averages, national or regional estimates of conversion factors can be used. But if the results are to be used to estimate farm-level characteristics a conversion should be worked out for every sample unit. In parts of Ethiopia the 'TIMAD' is a unit of area, said to be about one fifth or one sixth of a hectare. A study in 1985 used farmers' estimates of crop areas, but measured two plots from every farmer to work out a farmer conversion factor. The statistics for six districts (awrajas) are reproduced in Table 5.3.

The averages vary between awrajas, from the lowest of 0.14 (seven to the hectare) to 0.24 (nearly four per hectare). In this case a national average would distort the true areas in each awraja, but more important is the variation between farmers. Estimates range from one timad equal to 0.04 hectare, to one timad equal to 0.38 hectare. Clearly the use of an awraja average would misrepresent individual farmers. The study in question was designed to estimate cultivated areas and to produce farm budgets, so using a conversion factor for each farmer was correct.

An alternative to traditional units of area is a proxy variable, indicative of plot area. Some crops are cultivated in distinctive units, and counts of those units will approximate to land area. A good example is yam grown in mounds or heaps in parts of west Africa. In some communities the heaps are 'standardised' and are used as work units for casual labour. A count of units will approximate plot area, but conversion factors have to be calculated in the same way as for area measures.

Tape measurement

The simplest technique for measuring the area of an irregular polygon is to subdivide the plot into triangles and measure all the sides using a tape. The enumerator starts by drawing a sketch of the plot. Taking one particular point, triangles are constructed on the sketch. The sides of the plot are then measured, together with the distance between the fixed point and the vertices. If the shape of the plot includes a return angle, triangles can be drawn from two or more fixed points. To simplify computations the form should be designed for separate entry of each

Table 5.3 *Conversion of traditional area units to hectares*

	Hectares/timad				
Awraja	Min.	Max.	Mean	SD	No. farmers
Mota (1)	0.05	0.33	0.19	0.08	16
Mota (2)	0.15	0.30	0.22	0.05	10
Gonder Z. (1)	0.17	0.29	0.23	0.03	16
Gonder Z. (2)	0.14	0.38	0.24	0.06	16
Mendeyo	0.09	0.25	0.14	0.04	16
Genale	0.04	0.29	0.16	0.07	16

Source: Agricultural and Industrial Development Bank of Ethiopia
SD = standard deviation.

triangle. A completed example is shown in Figure 5.1. The area of each triangle is calculated by applying the formula

$$A = \sqrt{[s(s-a)(s-b)(s-c)]}$$

where *a, b, c* are the lengths of the three sides of the triangle in metres, and *s* is half the perimeter $s = (a + b + c)/2$. The area from this calculation will be in square metres and must be converted to hectares by dividing by 10 000 or as necessary into other units of area. The calculation is straightforward by hand, or easy to programme on a calculator or computer spreadsheet. If computations are to be done by hand it may be helpful to add a column on the form in which the value of half the perimeter can be recorded.

Tape measurement requires that the enumerator can see all the corners of the plot from one or two fixed points. This restricts the technique to low crops, or tall crops when immature or after harvest. Because the triangle sides are judged by eye, it is easy for careless fieldworkers to measure curved sides and get inaccurate results. For this same reason very large

Figure 5.1 Area measurement by triangles, plot sketch.

Triangle			Length of Sides (metres)			Area (ha)		
A	B	D	32	37	42	0.057	SUM ______	Plot area (ha)
B	C	D	22	33	37	0.036		
D	E	A	22	36	42	0.040		

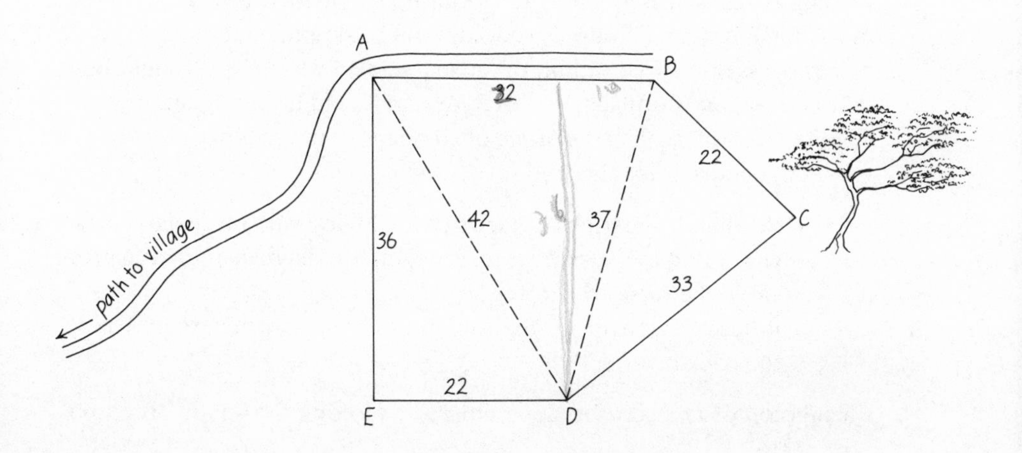

areas of land are not suited to this method. A further criticism is that there is no way of assessing errors of measurement. The main advantages are that the technique is simple to learn, easy to calculate, and does not need any special equipment. Although the calculation is simple, triangle sides tend to be easily confused, and the supervisor should check that each triangle is made up of the correct three sides.

Tape and compass measurement

For plots where a high standard of accuracy is required, or where access is restricted to the perimeter, tape and compass measurement of the plot boundary can be used. The equipment consists of a measuring tape (a surveyor's chain can be used) and a compass capable of accurate measurement of angles. A preferred design is a military-style hand-held prismatic compass. The enumerator visits the plot and after walking the perimeter with the farmer, to confirm the boundary, makes a sketch in which the sides of the plot are drawn and marked by letters of the alphabet. It is helpful for supervision if the sketch includes an indication of the path from the farm house, other prominent features, and the direction of magnetic north. Starting at one corner (A) the length of each side is measured in metres and the magnetic bearing in degrees from north is recorded (A–B). The enumerator will need an assistant for this, and the bearing is more accurately measured if the ends of each side are marked with a ranging pole. As the enumerator moves around the perimeter the back bearing of each side (B–A) should be recorded before measuring the next side. Each side is measured in turn and the distance, bearing and back-bearing recorded on the form. Bearings should be measured by standing back from each corner and sighting along the two ranging poles. On hilly land, sides can be subdivided to permit easier measurement. Irregular or curved sides can be simplified to straight lines by a process of compensation, or 'give and take' whereby a straight line between the ends of each side excludes and includes an equal area. A completed form, with an example of compensation, is shown below.

When the forms are checked several points should be covered:

- see that the bearings of each side are logical with respect to north
- check that the back bearing is within five degrees of the reciprocal of the forward bearing
- if necessary, average the forward bearing and the reciprocal of the back bearing

Calculation of area can be done either by plotting the shape on graph paper, and counting grid squares which lie inside the figure, or by means of a programme written for a calculator or computer. The FAO manual includes a number of algorithms.

Although all the sides of the plot are measured, unavoidable inaccuracies will result in the final point A′ not being exactly coterminus with the start point A. Most calculation algorithms present the Closing Error as being the distance A′–A expressed as a percentage of A–A′. It is common practice to accept errors not greater than 3%, but to resurvey plots which exceed this figure.

If a re-survey cannot be done there are surveying procedures to re-apportion the closing error but they are not generally used for plot measurements. In practice, after initial training, most enumerators can achieve a high standard of accuracy and errors tend to occur because of careless recording of measurements rather than poor accuracy. In such

Figure 5.2 Area measurement by tape and compass.

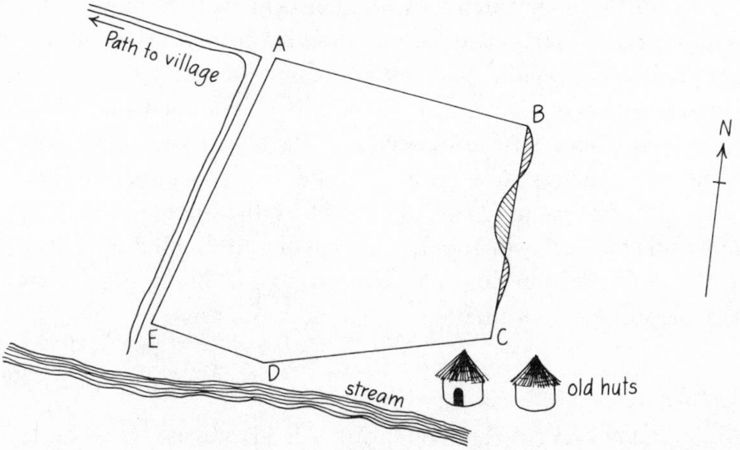

Side	Distance metres	Bearing fwd	Bearing back
A–B	22	097	277
B–C	18	175	365
C–D	20	243	063
D–E	10	295	115
E–F	25	007	187
F–G			
G–H			
H–I			

Plot area = 0·06 ha

Closing error = 0·82%

cases the closing errors are very large and conspicuous. Careful inspection of the survey details will usually lead to the error.

The tape and compass method can be used for plots of any size or shape, and because measurements are made at the perimeter it is not necessary to cross the plot, or even to be able to see across the plot. The main disadvantage is that area calculation is time-consuming by hand, or requires access to a calculator or computer. Occasionally an enumerator will have difficulty in reading the prismatic compass, but most can manage after a short training period. Some survey teams have used a system whereby plots are measured and calculated by the same enumerators. If poor accuracy is leading to frequent re-surveys a sharp enumerator will soon learn not to record the final side from field measurements, but to work out a distance and bearing which will nearly close the polygon, from a graph plot. Such initiative is impressive, but undesirable.

Concealment

The method of area measurement is likely to be less of a problem in some surveys than lack of cooperation by farmers. In many societies there are strong prejudices against discussing, or revealing, land details. Sometimes this is in order to conceal wealth or taxation liability, but often it is derived from distrust of outsiders. It is a well-known phenomenon that measures such as farm size have a tendency to be underreported, and to rise with increasing length of contact by the enumerator. To guard against this the surveyor should try to avoid attempting to collect farm size from limited contact or single visit surveys. In all surveys, try to cross-check the farmer's information:

- compare area reported with family size and likely consumption needs
- enumerate crop areas at farmers' fields and ask about ownership of adjacent land
- consult with traditional leaders, cooperative officials, extension agents and landowners, where appropriate
- inspect irrigation or settlement scheme records
- confront the farmer in front of other villagers about inconsistent information

Most important of all, make sure that the scheme of supervision covers important items such as land area while the survey is still in progress, and not after forms are returned to the office.

Summary

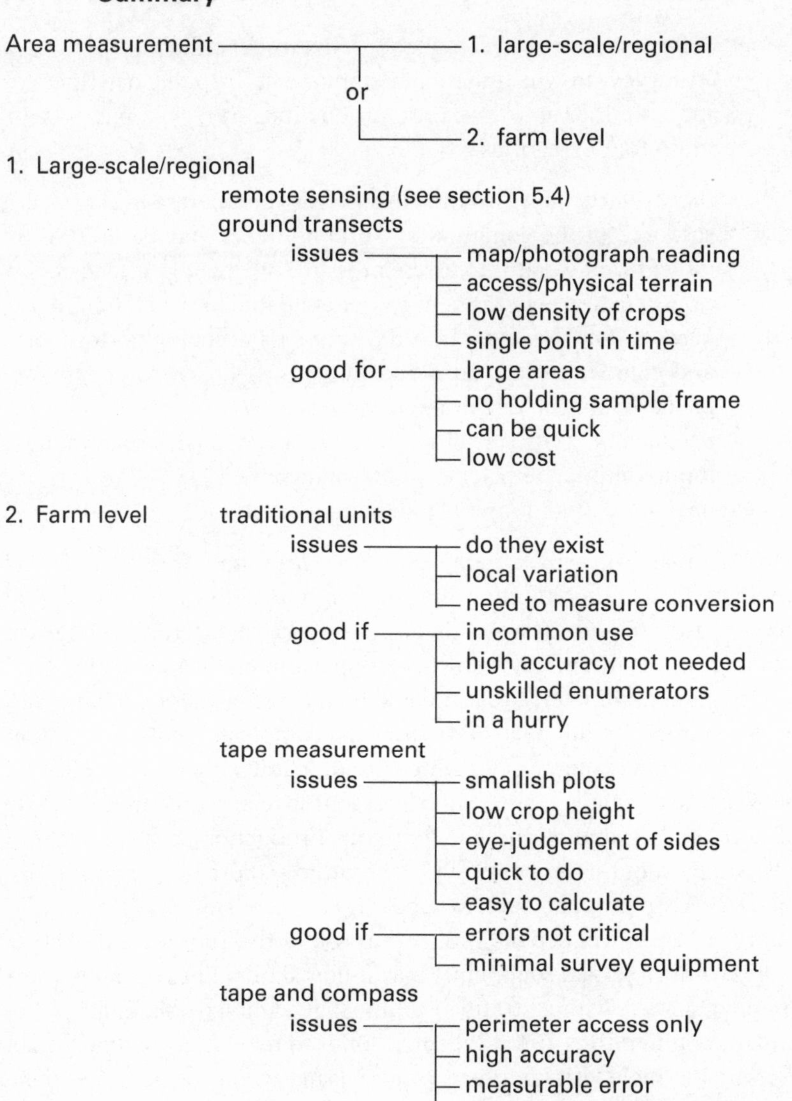
Area measurement
1. large-scale/regional
or
2. farm level
1. Large-scale/regional
remote sensing (see section 5.4)
ground transects
issues
map/photograph reading
access/physical terrain
low density of crops
single point in time
good for
large areas
no holding sample frame
can be quick
low cost
2. Farm level
traditional units
issues
do they exist
local variation
need to measure conversion
good if
in common use
high accuracy not needed
unskilled enumerators
in a hurry
tape measurement
issues
smallish plots
low crop height
eye-judgement of sides
quick to do
easy to calculate
good if
errors not critical
minimal survey equipment
tape and compass
issues
perimeter access only
high accuracy
measurable error
enumerator training
good if
need best accuracy
access to calculator/computer

5.3 Livestock data

Information is collected about livestock either as part of a household survey, in which the main emphasis is cropping activities, or for specific enquiries into livestock production. Survey topics can be classified into four categories:

- *the inventory* in which the purpose is to enumerate livestock as a resource of the community. Total numbers may be used as an indicator of wealth, to assess pressure on grazing or feedstocks, or as a measure of draught power available in the community
- *herd or flock structure* in order to assess breeding performance and characteristics, sales and purchases, life-cycle events and farmer management strategies
- *production* to determine physical productivity parameters, input–output coefficients and farm income
- *health* investigations by epidemiological studies

In the main, livestock data collection does not present different problems from other farming activities although the long production cycle of cattle and other large animals necessitates a different strategy for enumeration. Rather than following a single animal through a life cycle, development parameters are estimated by recording historical life cycles for a number of animals. Two aspects complicate data collection: livestock-owning communities which are nomadic or semi-nomadic and follow a transhumant grazing pattern present special problems for sample selection and enumeration; and continous production activities such as milk or eggs require frequent detailed recording (there is a similarity with some tree-crop products such as rubber).

Survey design will depend on the purpose of the enquiry. Production studies within a settled community may follow similar lines of enquiry to a cropping survey, with a stratified multi-stage sample design and static, resident enumerators. But if the intention is to take an inventory from a large area a single visit interview, or aerial survey may be used.[2] Specific data collection problems are listed below.

Inventory

In all farming communities concealment of land, crops and animals is a constant problem for surveys. Land and crops have the advantage of being immobile, but animals can be walked away. In the face of determined opposition it may be impossible to obtain accurate numbers from farmer interviews even when supported by observation. If the main purpose of the survey is to count animals, a combined approach, making use of household interviews and ground or aerial transects should be considered. With good cooperation herd details can be recorded by age and sex.

Herd structure

Herd structure analysis has been the most widely-used approach to studies of traditional livestock. For development planning a minimum set of data would include total numbers, age and sex structure, fertility and mortality. To collect such data is demanding on resources. For a moderate sample of herds several thousand animals have to be examined individually by skilled enumerators who are trained to age them by looking at their teeth, and record individual histories from their owners. If a single-visit enumeration is used, births and deaths over the last year should also be recorded. This approach records the situation at a single point in time. For long-term planning it needs to be repeated annually with the same herds, preferably with tagged cattle. The volume of work is a serious obstacle to this kind of study.

Ideas have been put forward to simplify data collection. First, reduce the need for skilled manpower to make use of farmers' own terminology for animal age, sex and history, so that data can be collected by less trained enumerators. Second, shift the focus of enquiry away from the herd, which risks the concealment problem noted above, to the individual animal. Individual female histories are known to livestock owners and can be used to model herd profiles by computer, using a range of estimates.

Nomadic pastoralists

Data collection from nomadic pastoralists is complicated by the shifting pattern of existence and the remote, demanding environment in which they live. It is difficult to find enumerators willing to live and work in such conditions. An approach which has been used sucessfully is to gather data on total numbers and distribution of animals from aerial transects. Aerial surveys have the advantage of being able to cover large areas quickly, and at relatively low cost per square kilometre, but the observations are of a single point in time and must be supported by other information before conclusions are drawn about movement patterns. The aerial survey can be combined with selective visits to herds, chosen from the aerial observations, for herd or animal histories. Aerial surveys are mentioned at the end of this chapter.

Production

Basic production data involve a combination of herd structure, vital events (births, deaths) sales, slaughtering and rates of weight gain for young animals. There is little alternative to detailed enquiry using either the herd structure approach, or individual animal profiles plus regular weighing. A large number of animals will need to be studied to obtain a range of observations.

Output

There is no short cut to frequent data collection of continuous production such as milk and eggs. From a sampling point of view the recording frequency should be chosen to suit the accuracy and precision required from the survey. The choice is between daily visits, frequent less-than-daily visits, or a single point estimate. A single point estimate by memory recall will tend to produce a long-run average estimate of output and may not reflect the specific period accurately. Frequent visit recording suffers from similar problems to those described for labour utilisation in Chapter 6, but can be an efficient compromise, and is especially useful if the farmer is able to keep output records. For the highest accuracy, daily visits are unavoidable.

5.4 Remote sensing

Remote sensing is included as a short section because it has featured as an optional approach for a number of data items, especially crop areas and counts of livestock. Remote sensing is a generic term which includes aerial photography and satellite imagery, both in the visible and non-visible spectra.

Satellite imagery is well known, and most agriculturalists will be familiar with images produced by Landsat. With earlier generations of satellites the main constraint to use of the data for farm studies has been the degree of resolution. The Landsat pixel (smallest image unit) represents 80 metres by 80 metres, which is considerably larger than the plot size used in many smallholder farming systems. More recent satellites, notably the French SPOT, operate at a lower level, around 20 metres square. At this level good regional and national estimates of land-use characteristics are possible.

If conventional ground-based surveys are designed to produce regional land-use estimates, satellite imagery offers an alternative approach, or at least an independent source for comparison. But most ground-based surveys are concerned with economic units, such as the farm and the community. At current resolution available for non-military uses, satellite imagery is unlikely to be a serious alternative to ground surveys.

By comparison, aerial photography does offer sufficient detail to be used alongside questionnaire surveys. Conventional aerial photography will permit the identification of farmers' plots and if it is up-to-date could be used for the measurement of land areas. But conventional aerial surveys are expensive and unless the photography happens to coincide with the survey year, it is unlikely that photographs would be available. Recent photography is still valuable as a record of settlement location and land-use patterns, and can be used as a sample frame, but the high cost inhibits frequent use.

An attractive development in the collection of data from the air has been the use of low level aerial observation and photography. The costs are lower for several reasons. The aircraft used can be smaller and cheaper to operate; photography is carried out to a lower technical standard (image quality and overlap between photographs) because the images are not intended for stereoscopic viewing and map-making; and photography can be done with small-format cameras (35 mm and 70 mm).

Aerial observation has been used extensively for counts of wildlife and livestock. The technique is to fly at a predetermined height, usually not more than 1000 ft. As the aircraft proceeds along a flight line, observers sitting on each side record objects on the ground from a strip defined by a viewing angle from the aircraft window. Small herds or settlement groups are counted; larger ones are counted and photographed with a hand-held camera to cross check numbers during analysis. The sample strip can be calculated by knowing aircraft height and the angle of view. Observations can either be made by flying along a grid network, in effect giving a systematic sample, or by a random set of flight lines chosen with reference to the location of points of interest on the ground – a stratified random sample.

This technique has been extended by the use of vertical photography. In an aerial variation of the ground transect survey photographs are taken at regular intervals along grid transects. Analysis of the land use on each photograph permits an estimate of land areas to be made.

In cost the technique is comparable with satellite imagery, and much cheaper than conventional aerial photography. It is attractive for survey data from large areas of land with low settlement density, where ground-based studies are difficult to manage. And if a systematic sampling procedure is used the data have the advantage of being gathered from throughout the survey area, unlike ground-base studies which tend to be clustered at relatively few sites.

Notes and references

1. There is a wide variation in the definitions used for holding, field and plot. A discussion follows in section 5.2. See also the glossary.
2. Readers interested in livestock studies can consult Putt *et al.* (1988), Jahnke (1982), and Swift (1981).

6

Data collection: Methods II

6.1 Unit of study

Many farm surveys are prompted by the data requirements for planning, or evaluating the success of, development projects which involve technical changes in crop or animal production. Information is sought about the use of fertiliser and crop protection chemicals; and decisions made by farmers about crop varieties, planting dates, and planting densities.

The data in this category can be collected by a mixture of interview and observation. One of the main decisions facing the survey designer is whether to rely on memory recall to quantify the response, or to try for direct observation. This will affect the visit frequency, and therefore the cost of the survey, and must be suited to the desired precision and accuracy of the results. Inputs are considered first, followed by management practices.

The simplest way of estimating the use of an input is by *unquantified recall* interview with the household head or farm operator. The question can be as simple as 'Did you use product X this season?' This can be done with minimal enumeration resources, because no technical skills are involved, and for an exploratory survey can give a swift indication from a large sample. But it is a simple matter to extend the question to *quantified recall* by asking about the quantity of input purchased or used. Recall of a purchased item from the recent past is likely to be accurate because it is unlikely to be a routine event, and a farmer could be expected to remember correctly.

Wherever possible, the enumerator should ask to see the bag or container in which the input was supplied, or other proof of purchase, in order to confirm the interview response. A clear distinction should be made between the quantity purchased and the quantity used. In some instances the survey may require both estimates to be collected. If it is the quantity used that is to be measured a decision must be made to record the quantity either for the whole farm, or by specific crops or for individual plots. This will depend on the use to which the data are to be put. If the purpose is to determine which crops the farmer chooses to apply inputs to, a simple crop listing will suffice. But if the aim is to compute an application rate, and try to relate crop output to the use of inputs, plot-specific measurements must be made.

The plot definition can be an obstacle to accurate quantification of inputs in the same way as described for crop production. If the farmer applies an input to one crop in a field, which is subsequently divided into plots by the addition of a second crop on part of the field, the farmer will not know the application to each plot separately. To assist the enumerator, provision should be made for this information to be recorded for the whole field, to be allocated to plots in proportion to plot area at the survey office. In all data collection at the plot level it really is important for the enumerator to conduct the interview at the plot, so that visual observation can be used to prompt the farmer.

Although they are often given less prominence than chemical inputs, farmers' decisions about planting dates, weeding and plant densities are equally important elements of crop improvement. And for many planning surveys they are essential to an understanding of the existing farming system. Three collection methods can be used: *farmer recall with limited observation*, *direct observation with limited skills*, and *direct observation with technical knowledge*.

Some data items generally come direct from farmer interview with limited opportunity for observation. The main items here are the timings of activities; land preparation, planting, transplanting, thinning/supplying, application of inputs, weeding, and harvesting. They are normally recorded from interview with the farmer because the enumerator is unable to visit plots frequently enough to observe each activity. However, many of the operations leave visible marks on the plot which should stimulate the enumerator to check with the farmer for information, or to crosscheck information already given. Interviews should take place at the plot to avoid confusion and for the visual prompting.

A further set of data items can be collected by direct observation, for which basic skills are needed by the enumerator: land type (dry land, wet land, irrigated land), type of land preparation (flat, ridged, raised beds, etc.), the method of land preparation, where the crop is planted (on the flat, on ridges, in furrows, etc.) and the stand or plant density. All the information can be obtained from direct observation at the plot, plus a farmer interview to confirm details of methods or suchlike.

More advanced characteristics, such as the incidence of specific pests and diseases, may be extremely valuable to the recipients of the survey report but require more careful training of enumerators and a good understanding of farmers' own description of the infestation. This kind of detailed information is often reserved for informal surveys conducted by crop and animal scientists, rather than enumerator surveys, because of difficulties in ensuring consistency of understanding by farmers and interpretation by the enumerator. There are also problems in quantifying infestations over the plot in a way that is comparable between farmers.

6.2 Population

Purpose

Population estimates are common to many rural surveys. Most survey work is based on household samples or samples of household members; an estimate of the survey population is necessary to calculate population statistics, and investment benefits are estimated from project populations. The spatial distribution of households is used to plan road networks, water supplies and the provision of input supply centres. Nationally, population estimates are used to plan the growth in the labour force and provision of government services. Regional population may be a component of calculations of government expenditure allocation and revenue generation. Population statistics are important, and because of their importance, politically sensitive in many nations.

In rural surveys the task is usually to obtain a population total for the calculation of survey estimates, and an estimation of household size and population age–sex structure to analyse the rural community (see Casley and Lury, 1981, Chapter 12).

Methods

The most rigorous and formal method of estimating a population is a full census, where every household is visited by an enumerator and questioned about the household members. This is rarely an option for a small-scale rural survey. But the techniques are similar to a sample enquiry, as discussed below, and the main difference relates to the scale of operations and consequent requirements for logistics and organisation. The formal census is not explored further here.

Possibly the most common requirement is related to the needs of the sampling frame. A typical situation is a two-stage frame where a total population is needed for estimation, a size estimate is needed for a PPS sample selection of villages at the first stage, and a listing of households is made for a SRS selection at the second stage. A cost-effective way of approaching this problem makes use of three stages. First, use existing data, updated historical estimates or estimates from the communities themselves for the first-stage selection. At the second stage make a listing of all households in the selected villages for the SRS selection. From those households chosen for the final sample, record household size and the age–sex structure. The household listings at the second stage can then be used to update the original estimates and calculate a correction factor which can be applied throughout the population to update the population. Household size from the survey sample can be used to calculate total population or total number of households as necessary.

If historical estimates are not available, and circumstances prevent the use of informal estimates by community leaders, it may be possible to work from proxy data such as dwelling counts from aerial photographs. The procedure is to use these figures for first stage selection, and apply a correction factor based on a household count from sampled settlements at the second stage. Care must be taken that the aerial photograph is not long out of date, or that settlement size and location have not changed significantly since the photography.

Collection issues

When the unit of study for the sample is based on the human population a precise definition must be given for the survey. The most common example, used casually throughout this book, is the household – a concept that is comfortable and recognisable in Europe and North America, but which frequently does not easily transfer to smallholder agrarian communities.

A common definition used in a number of African countries is 'a group of people who eat from a common pot. They usually share dwelling houses and may cultivate the same lands. They recognise the authority of one person, the pot head or household head who is the ultimate decision maker for the household.'

This definition hinges on four features: consumption of food, dwellings, cultivation of land, and acceptance of a common authority. How should it be applied if one or more of those features does not hold true? A survey in West Africa where a similar definition was being used found a number of simple exceptions. An old man lives in a compound with his brother, one adult son and their kin. They eat separately yet their grain is stored in one common granary and they farm the land as a single production unit. In another instance a wealthy farmer lives and cooks in one compound. His adult sons live and cook in another compound. They farm 4 hectares jointly as a single production unit, but the sons also work individually on other fields from which they benefit directly and personally. How many households are there according to the definition in these two examples? The problem is that dwellings are not necessarily synonymous with household, and production units (sometimes based on age–sex groupings) may cut across both family and consumption bonds.

Definitions should be worked out in collaboration with people who are knowledgable about the community being studied. Whichever definitions are adopted exceptions will be found and careful training is necessary to ensure that enumerators bring exceptions to the surveyor's attention rather than disguise the problem (Casley and Lury 1981:186).

When recording household members a convention must be adopted to record either the *de facto* population, all those resident at the household when the enquiry takes place (a common benchmark is to ask for those who slept at the dwelling on the night before the interview), or the *de jure* population, all those whose family relationship gives them a right to be a member of the household, but who are not necessarily living there. They may be working, or at school away from home. For surveys of agricultural production the *de facto* population is usually of interest. Variations in the period of absence can be used to modify the definition to include all those who are absent but due to return within a short period of time. It is also important to time the household population estimate to coincide with the agricultural season, so that household size reflects the domestic labour force. Some surveys which last the whole season update household population at every interview or at defined points such as land preparation, mid-season and harvest.

In some societies household size can be quite large and a count of 10 to 20 people is not uncommon. In such circumstances it is easy for some people to be overlooked. Typical missing categories are babies on their mother's back, young unmarried men and women in seclusion. If cultural mores permit, the enumerator should try to see every member of the household during the interviews.

For middle-aged and elderly persons, age may not be known. A helpful prompt is to construct a calendar of events which charts events in the history of the community and nation. Age can be approximated by a combination of remembered events and relationships within the family. It is usually a good idea to record the exact age for each person even if subsequent analysis only requires a division into broad categories. When ages of other family members are being reported by the household head there is the inevitable problem of inaccuracy. Age heaping is where ages in rounded numbers such as 20, 25, 30, 35 etc are more commonly reported than the years in between. There is evidence that the age of women is sometimes moved into the attractive child-bearing range of 15 to 40 years, or that the age is adjusted to fit conventional characteristics, so if a woman has passed the menopause she must be over 45. Enumerators should remember to record the age of babies less than one year as 0, not 1, and for frequency distributions it is conventional to form classes from the age groups 0–4, 5–9, 10–14, etc. These classes are directly comparable with international census data.

Although population is collected in most rural surveys it is not always actually necessary, and there is a tendency to overload questions about members of the family. Typical additions are questions about education, literacy, present occupation, settlement migration, labour migration, farm labour activities, and availability for farm work. This tendency is especially prevalent when the data are collected in a simple row/column table, as illustrated in Example 7.9 in the next chapter. There is a danger that questions which in fact require very careful enumeration are treated as a simple multiple choice option, with resulting poor data quality. If other information has to be obtained it is better treated as a separate set of questions rather than an addition to the household age–sex record.

6.3 Labour

Purpose

The uses of labour data fall into three broad groups. First, is the construction of enterprise and farm models for which labour coefficients by crop, activity and season are essential components. Such models may be used to describe existing farm circumstances or to illustrate the effects of proposed changes. Second, is analysis of labour utilisation and productivity both for the household as a production unit and at a more aggregated level for sector studies of productivity. Third, is the study of labour force components to determine changes in the labour force and participation in farm or non-farm activities. In this section our concern is with farm operations.[1]

Data collection issues

Farm activities are a good example of activities which produce data which are repetitive, continuous and detailed. To minimise sampling error from such data sample size should be set as large as possible. But the type of data are such that great care must be taken over measurement and recording. For labour utilisation, measurement error is related to the type of data collection (direct observation or memory recall) and the time which has elapsed since the activity being recorded took place. To minimise measurement (non-sampling) error sample size should be reduced. There is a trade-off between the size of sample and the frequency of data collection. The most intensive frequency would be daily recording. If the trade-off between sample size and periodicity of interview was not affected by memory recall, then a two-day period could replace daily recording, with double the sample size, and a four-day period with four times the sample size, etc. But as we discuss below, memory recall becomes an important consideration for both long and short period intervals.

A second aspect of labour data concerns the definitions of operations. Detailed labour studies have shown that work input varies by type of worker (age and sex) and motivation (family, contract, social work-group) as well as by physical factors. Thus the labour required to form ridges for planting will vary according to whether the land has previously been cultivated or not, the labour for weeding will vary between a first or subsequent round of the activity, and the time actually taken will depend on the motivation of the person doing the work.

Not all households or plots will experience the same labour activities and great care is needed in reporting data to indicate the number of observations of each class of labour and activity. Table 6.1 illustrates a labour data report.

A third general consideration is that farm work will usually involve several members of the household. In some cultures, specific age–sex groups will have responsibility for different operations, such as land preparation by adult men, weeding by women or herding small stock by young boys. If the survey interviews are conducted by the household head he or she will not necessarily know the actual work done by other members of the family. If social customs permit, enquiries should be made with the individuals concerned.

Table 6.1 *Reporting labour data for crop enterprises*

Crop mixture	Yield rate (kg/ha)	Stands (ha)		
Sorghum	423	9795	Sample size:	184 plots
Millet	205	6061	Mean plot size:	1.12 ha

	hrs/ha		hrs/ha
Land clearing:		Harvesting:	
Adult male	7.22	Adult male	60.69
Adult female	0.54	Adult female	2.80
Child	0.37	Child	12.79
Family total	8.12	Family total	76.29
Hired	4.78	Hired	38.89
52% Total	12.90	100% Total	115.17
Land Preparation:		Post Harvest:	
Adult male	11.88	Adult male	1.28
Adult female	1.43	Adult female	0.54
Child	2.51	Child	0.64
Family total	15.82	Family total	2.46
Hired	5.65	Hired	0.95
53% Total	21.46	5% Total	3.42
Planting:		Transporting inputs:	
Adult male	18.08	Adult male	0.77
Adult female	1.73	Adult female	0.00
Child	2.14	Child	0.02
Family total	21.95	Family total	0.79
Hired	1.20	Hired	0.00
100% Total	23.15	7% Total	0.79
Post Planting:		Transport outputs & sales:	
Adult male	103.63	Adult male	3.52
Adult female	4.53	Adult female	0.29
Child	29.82	Child	1.18
Family total	137.97	Family total	4.99
Hired	57.90	Hired	6.38
99% Total	195.87	50% Total	11.37
Fert. & chem. application:		Total labour	
Adult male	6.25	99% Adult male	213
Adult female	0.21	15% Adult female	12
Child	1.10	41% Child	50
Family total	7.57	Family total	276
Hired	1.03	52% Hired	117
47% Total	8.60	Total	393

Hours per hectare are averages over the whole sample size. Percentages show the proportion of plots on which an activity or a type of labour was reported. Thus fertiliser and chemical application was reported on 47% of plots. Child labour was reported on 41% of plots.

Methods

Two main types of data collection methodology have been used: direct observation and memory recall.

Work measurement

Work measurement, or more correctly time study, involves the detailed timing of labour activities by direct observation. Two variants are possible. To follow specific household members throughout the day, and record their activities, or to observe an activity by a worker or group of workers for a specific time period and record the work done at the end of the period. Clearly a considerable effort would be required to conduct this operation on a large scale, at considerable cost, but this could be justified if the data were of exceptional quality.

A number of problems are thought to affect this approach. There is a well-documented tendency (the Hawthorne effect) for people to work harder when being observed. The possibility exists that this would be affected by the length of observation, and by the experience of the workers with previous observation. Length of observation is also important in consideration of rest periods, which affect the rate of work, and are more likely to be observed over longer recording periods. These effects would tend to overestimate the rate of work and underestimate total hours.

Other problems concern the difficulty of scaling-up results from a plot observation to a hectare basis. Treatment of overhead time, such as travelling from the house to the field, which is as high for a small plot as for a large one, requires care. Different lengths of jobs pose problems. It may not be possible to observe the actual work done during a short observation of lengthy activities, such as tending cattle.

Work measurement requires fewer data than do frequent-visit surveys for similar levels of accuracy and the data are therefore more rapidly and cheaply obtained and processed. Overall, the technique may be attractive for some detailed studies but the organisational complexity is undesirable for a large-scale enquiry. The results obtained are thought to underestimate labour utilisation, although comparison with frequent-visit surveys is complicated by the problems of memory bias, which also affect estimates.

Memory recall

Labour data, like other survey information, can be collected by frequent- or limited-visit regimes. Both systems are affected by the ability of respondents to recall information. Memory recall is a function of the nature of the activity to be remembered, the reference period during which the activity took place, and the recall period or time which has elapsed since the reference period. Infrequent events are more likely to be remembered individually than frequent events. Regular events are unlikely to be remembered individually but their regularity may provide a pattern which will aid recall as to whether the events are frequent or infrequent. In other words, activities such as the use of hired labour are more likely to be remembered than family labour, and the regularity of seasonal events will act as a guide, but isolated events such as the purchase of fertiliser are likely to be remembered more accurately than labour data.

The accuracy of a respondent's memory is clearly associated with the length of the reference period and time of recall. Zarkovitch (1966) and Lipton and Moore (1972) both refer to a process of memory fading. The longer the reference period and the longer the time between the reference period and date of recall the less accurate will be the information. The implication is that both periods should be kept as small as possible. But this problem can be mitigated to some extent by the choice of the reference period.

It is important for the respondent to be able to locate the start and finish of the period. If this is not possible it is considered likely that some events will be transferred into the period from adjacent periods and other events will be transferred out. This transference will be notable with respect to events occuring close to the two ends of the period. Furthermore, Zarkovitch has argued that this transference is particularly frequent with short reference periods, because respondents try to recall individual events, whereas on longer reference periods fewer transfers are made and the recall tends towards a form of long-term average.

The two main choices, between frequent visits, daily or twice-weekly, and limited or infrequent visits, both involve an apparent disadvantage. If highly accurate labour data are needed, there would seem to be no alternative to daily recording, because overestimation is likely to occur with visits as frequent as weekly. But the cost would clearly be high for all but small samples. If indicative data, representative of long-term averages, will suffice, a better plan would be to make a limited number of visits, timed either to recall the whole season, or at specific points such as after planting, after weeding, etc., with the aim of recording each stage in the seasonal cycle.

6.4 Household resources

Many surveys include an inventory of domestic possessions and farm equipment: vehicles, electrical goods, type of furnishings, type of house construction, ploughs, milling machinery and tractors. Generally the aim is to build up a picture of the distribution of economic resources, with the intention of reporting changes over a period of time, or of classifying the sample by wealth categories as part of the analysis. As long as care is taken for the enumerator to try to see, count or measure each item, and not rely solely on an interview response, the subject does not provide any particular problems for data collection.

The main issue is how to draw up a suitable list of items. If change is to be measured it may not be obvious how the survey population would react to changes in income, and therefore which items will show a response. In some cultures clothing and personal items such as watches, radios and motorcycles are early evidence of rising incomes. Elsewhere, education for children may be a higher priority. A common survey approach is a long list of possessions, some trivial, including items such as cupboards and matresses, which are an irritation to both the enumerator and respondent and as a result are badly reported.

For good results the list should be based on careful exploratory investigations, kept short, and precisely defined so the enumerator does not have to make judgements between different types of articles. Resource data are personal to the household, and wherever possible its collection should be deferred until the enumerator has built up a relationship with the respondent. Because of this it is not ideal for collection from a single-visit survey.

6.5 Income, expenditure and family consumption

Purpose

A frequent demand is for rural surveys to be extended beyond agricultural information, to economic and social characteristics of the household. The demand is reasonable on two counts. First, the growth of public-sector investment directed towards poverty-related issues is based on concerns that some groups in society are relatively disadvantaged. Unless data exist at a micro level it is not possible to plan how to identify and target people in poverty, and subsequently to follow-up investments with evaluation of their achievements.

The absence of detailed household-level data has rarely prevented project investments, especially in the rural sector. But paradoxically, the realisation that in addition to projects, macroeconomic interventions such as price setting, fiscal policy and exchange controls need to be adjusted to stimulate efficient use of resources, has led to increasing demand for household-level data. Governments need household-level data in order to monitor the effects of policies such as structural adjustment, on selected socio-economic groups in the population.

The second reason for a continuing demand for household surveys concerns the dominant role still played by the household in many countries. For large proportions of the population the household is a coherent social grouping, a focal point for social and economic activities and decisions. Increasingly, urbanisation and the expansion of wage employment will bring changes. Even now, the definition of a household may present a problem (see section 6.2) especially where members are labour migrants on an uncertain basis. But the household remains a workable and convenient unit of study for a wide range of topics.

Prominent amongst the list of topics are estimates of poverty, employment, migration, access to social services and food security. These issues can be broken down to fundamental measures of income, expenditure and consumption.

Data collection

The most important feature about these data is that with the exception of wage-earning or similar transactions with documented records, direct measurement or observation is virtually impossible. Personal interviews are the only source of information, and the problems are very similar to those for labour data:

- estimates by memory recall require carefully defined, closed reference periods[2]
- income, expenditure and consumption events are mostly regular and routine, with occasional items which are uniquely memorable
- repeated, frequent visits to collect data are probably more accurate, but can be irritating for the respondent
- single visit interviews which use long-period recall are prone to accidental distortions, such as transfers of events into or out of the reference period

- the household head will not necessarily know details for all household members

Furthermore, if the option of a long-period recall is taken, to collect data for, say, one year, care must be exercised that the respondent is asked in units and quantities that are likely to be known. Because annual income, expenditure and consumption are made up of numerous small events the respondent will only know the quantities in those units. As a general principle questions should be built up from basic components:

- value of crop output, crop by crop
- income from livestock by class of animal and product
- expenditure prompted by crop, animal and seasonal operations
- off-farm income, person by person
- sources of off-farm income prompted against a checklist
- family consumption prompted by daily routine and a checklist of items

In this way it is easier for the interviewer to spot incomplete or inaccurate replies and the style of the interview does not highlight total figures, which are perhaps likely to be sensitive and prone to distortion. For each item the survey form should include provision for both a physical quantity and a value assessed by the respondent.

In the context of project-related surveys for planning or evaluation, household surveys are a major undertaking, and should not be undertaken unless there is clearly an adequate provision of skills and resources. Properly financed at a national level they are a vital component of a government information system.

Good progress has been made by the UN with the Household Survey Capability Program (UNHSCP), which began in the late 1970s, and more recently by the World Bank's Living Standards Measurement Study (Glewwe, 1990) and Social Dimensions of Adjustment data collection activities (Grootaert and Marchant, 1991). The publications referenced describe the statistical and practical issues of running household surveys, and are valuable reading for anyone considering this type of investigation.

All this notwithstanding, it is prudent for the survey designer to question the wisdom of even trying to collect income, expenditure and consumption data, before embarking on design and exploratory surveys. Unless very high standards of enquiry are achieved the results are likely to be unreliable, and potentially damaging if users of the data are not aware of their shortcomings. An alternative approach is to avoid the problem of measuring total income or expenditure by concentrating on physical production, which can then be modelled using price and marketing data. Proxy measures of wealth, and access to or participation in social activities such as education, may convey sufficient information about economic well-being. If a survey is unavoidable we suggest that a small (case) study of a few households under good supervision will produce more reliable and usable data than a large-scale sample survey. Expenditure data are likely to prove more reliable than income data.

6.6 Price data

Price data are used for two main purposes: as an indicator of market and general economic trends; and to value goods and other economic data. The familiarity of the term disguises subtleties which affect data collection.

The first concerns definitions. There is no such thing as price, there are several different prices. For farm production we may distinguish between prices at the field, farm gate, wholesale and retail. Each stage is separated by costs; of transport, storage and treatment; by the influence of bulking in transactions and for transport; and by market knowledge on the part of sellers and buyers, when they bargain. It must be clear at which transaction price information is to be gathered.

Second is the state of the items being priced. Threshed maize differs from threshed and cleaned maize, basmati rice differs from blue bonnet rice, seed millet is not the same as millet for consumption. Survey forms must specify precisely the state and condition of the items being collected.

Data collection is mostly done in one of two alternative ways. Prices can be collected by recall from traders or farmers and other household members, or they can be collected from direct observation and measurement at markets. Prices recalled by memory suffer from the usual problems of memory recall: they may not coincide with the actual period of enquiry; recall may be influenced by isolated events; and the data cannot be checked against actual measurements. If data are collected in this way, a useful check can be made against income and expenditure data, which, if both physical and financial data are recorded, implicitly contain price valuations.

Market price surveys are a popular approach to price information. The most frequently encountered method is to sample traders at a market site.

From each trader a sample of produce is weighed on accurate (kitchen-style) scales and recorded against the sale price. Variations can be incorporated to sample traders in certain goods, sample in proportion to total numbers of traders, and sample during the course of the marketing period.

The problem with this approach concerns the way a price is determined. In some countries fixed or guide prices are used at formal markets. But elsewhere, and in free market bargaining, a price is only agreed after discussion between a genuine seller and a genuine buyer. In other words, an investigator will not collect actual prices unless goods are bought. This argument is sometimes taken further to include the proviso that the buyer uses his or her own money, or else the bargaining does not carry any force. One way around this problem is to interview customers leaving a market after they have made their purchases.

Analysis of price data is more of a problem than collection. The difficulty is in generalising the results beyond the transactions which were recorded. Each price observation can be thought of as a point on a space–time surface. Generalisation beyond that point, to an average across a district, or an average over time is complicated by a lack of information. To average across space requires information about the quantities marketed at each price and location, so that a weighted mean can be computed. Market centres which attract many buyers and sellers may have lower prices than remote villages where few transactions take place. Clearly a simple average would distort the price in favour of the remote village, and mis-represent the prices received elsewhere. Equally, prices vary seasonally, and an average between periods of scarcity and periods of surplus would need to be weighted by the quantities marketed at those times. The only averages which are sensible are location-specific averages for short periods of time: market average per week etc.

The danger with average prices is in using them to value farm output. Small price changes can alter the value of output unrealistically. For example, perishable commodities such as fresh vegetables can attract high prices at town markets. But at the end of the day's selling some quantities may be left. Part will be eaten, the rest thrown away. How should the analyst value each part of the crop? The average traded price would clearly not be correct. Precise specification of quantities and disposal is essential for accurate results.

If the survey analysis makes significant use of price data it is good practice to recalculate budgets and other economic data using a variety of price levels, to illustrate the possible range of actual performance. This type of re-calculation has been made more straightforward by the use of spreadsheet software on microcomputers.

Notes

1. For discussion of the difficulties associated with the measurement of farm labour see Cleave (1974), Farrington (1975) and Coleman (1982). FAO (1978) includes a discussion of definitions and concepts. Collinson (1972) proposes a simplified form of memory recall based on seasonal operations.
2. Experimental work on the effect of varying recall periods on recall loss in Ghana showed an average loss of 2.9% for every day added to recall duration. See Scott and Amenuvegbe (1990).

7

Questionnaire design

A basic aim of data collection is to try to ensure that the information compiled for analysis is as near as possible to reality. The process of sampling respondents, questioning them, recording their answers, summarising those records, and processing and analysing the data involves a sequence of transfers. At every stage in this sequence there is a potential source of error, as illustrated in Chapter 4. Maintaining data quality is concerned with efforts to minimise errors and improve the accuracy of results. One major contributory element in this chain is the questionnaire design.

7.1 Questionnaire design

Before any attempt is made to list topics for a survey, serious consideration must be given about whether a form is necessary, or desirable. If at all possible the best option is to use existing data for part, if not all, of the study. If this option does not exist, then a survey must be considered. But not all enquiries are appropriate for questionnaires. Informal, exploratory surveys, or detailed follow-up to specific cases following a survey, do not require a questionnaire, indeed the very purpose of the enquiry may be to permit an open-ended discussion. They do, however, need a checklist of topics, which is the starting point for design of a questionnaire, and the process of making a checklist, laying it out for reference in the field and preparing a method for recording answers, follows closely the procedures with a questionnaire.

Within a questionnaire survey different classes of information may lend themselves to different styles of enquiry. Simple factual statements of farming practice may suit a formal questionnaire, but the accompanying exploration of marketing patterns, because they are more private to the household, may require a delicate, carefully-phrased approach which cannot be catered for by the questionnaire survey. In this case the enquiry needs to be split into more than one component.

If a questionnaire is to be used the design is critical because survey analysis depends on the completeness of the topics covered. A lack of care at the design and testing stage of a survey will result in questions which do not cover their topics in sufficient breadth or detail and therefore fail to provide the basis for analysis. Once the survey gets under way the form is the working tool of the enumerator. Problems and omissions will reduce his or her effectiveness and poor design may result in questions being left out or misunderstood. Error-free data transfer require clear, comprehensive questions, good enumeration, and clearly set out answers. Much of this process depends on good questionnaire design.

Principles

Some principles can be summarised to help with the design of forms. They stress the links between the form and the interview and between the form and data processing. With the exception of simple measurement the interview is the mechanism for collecting data but the form is the only lasting record of that interview. A good relaxed interview, with maximum cooperation from the respondent, can be wasted if the survey form cannot cater for the information given in reply. That information has then to be analysed, and in most instances verbal information will first have to be converted to numeric information, by a system of codes, before the data can be processed. The form must cater for coding and subsequent data entry for processing.

Six principles

- *Content*: include the *minimum number of topics* to meet the objectives. Normal practice is to ask about too many. To help the enumerator the interview should focus on items of direct and major interest. What does the survey want to find out; why is the information needed; from whom or where can it be obtained; and how are the topics to be questioned.
- *Time* for the interview must be kept reasonable: in general no more than one hour. This limits the number of questions. Also, the more detailed or thought provoking the topics, the fewer questions there should be.

- The questionnaire should be *easy to use* as an interview guide for the enumerator and as an instrument for recording answers.
- The questionnaire should be *self-contained*, with identification of the enumerator, respondent, date of interview and any other reference information such as field details included.
- *Coding* for analysis should be done directly *on the form*, preferably alongside the verbal response to each question.
- *Smart presentation*: careful thought should be given to the quality of paper, the size of the sheets used, the clarity of printing and presentation and the spaces provided for recording answers.

The process of design is creative and surveyors develop strong preferences for particular styles of layout and phraseology. In this chapter we try to avoid prescription, but attempt to identify good and bad features of design, around which a form can be modelled.

A typical sequence of activities to design a form would follow this pattern:

- Draw up a list of question topics from a mixture of theoretical models, empirical information, research evidence and terms of reference for the study
- For each topic phrase the specific information required
- List them in a logical order, following either a chronological or a sequential pattern
- Decide for each question how to record the interview response
- Make a first draft layout on the style of paper to be used
- Test the design on model respondents
- Prepare a pilot draft for a pilot or test survey
- Modify the form from the results of the test
- Finalise the design and layout

Review as many times as possible the number of questions finally listed. Most surveys contain too many questions; on well-worn topics, or as shopping lists, or 'just in case'. The just-in-case problem is a real one. An inexperienced person asked to conduct a survey dreads the idea that during analysis a key explanatory variable will be found missing, so family size and structure creeps into every agronomic form, the education experience of the household head accompanies each enquiry into plant spacing, etc. The dilemma is real, it takes great professional confidence to exclude questions. And after all, the main cost in a survey is in reaching a sample unit, a few extra questions take little more time. Perhaps the best that can be hoped for is to remember 'if in doubt, leave it out' – for analysis if not in the field.

Form design is largely a compromise between opposing criteria – the ease, speed and accuracy with which the questionnaire can be completed in the field, versus the ease, speed and accuracy with which information from the questionnaire can be processed for analysis. The skill of the designer is to give equal attention to both aspects.

Data collection is enhanced if the questionnaire is:

- laid out with the questions fully specified
- completed in words rather than codes
- laid out according to the logic of the subjects covered in the interview

Data processing is facilitated if the questionnaire is:

- laid out mostly in blocks of figures, with occasional key words to indicate the nature of the data items
- completed directly in codes rather than in answers expressed in words
- laid out according to the logic of the data processing procedure

The procedure of designing a form has four key elements:

- the sequence, detail and content of the *questions*
- the method of recording *answers*
- *coding* for data processing
- the *layout* of the pages

This chapter looks first at questions and identifies specific problems with phraseology. The main scope for alternative form design concerns the way in which answers are recorded. These are dealt with next, followed by a section about coding. Last is a look at options for laying out forms (Casley and Lury, 1981: 91).

Summary

Look for alternatives
- secondary data
- checklist
- formal questionnaire

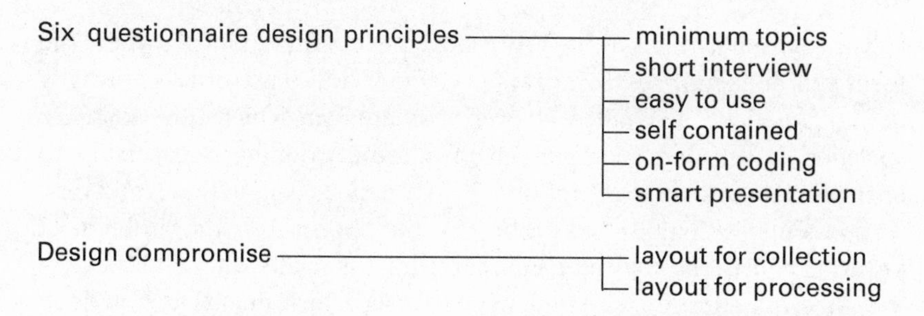

7.2 Questions

Question layout

As a general principle, questions should be presented in a logical order, designed to follow a natural sequence, like a crop season or an administrative process such as taking credit. Alternatively, if the questions do not reflect a sequential set of topics but include potentially sensitive or embarrassing items, the order should be arranged to ensure that the least sensitive items come first in the hope of relaxing the respondent and gaining his confidence and cooperation. More sensitive topics can be worked into the content after preliminary routine questions. Although this sounds straightforward, the notion of a sensitive topic differs markedly between cultures. Income and wealth-related subjects may be prone to vague answers by most cultures, but less obvious topics such as crop storage, mention of the last-born child's name, or the age of elderly people may be socially unacceptable. Careful review of proposed questions by people who are familiar with the culture of the people being surveyed is essential here. (See Sudman and Bradburn, 1982.)

Four basic alternatives are found in the layout of questions.

- A verbatim listing of every question, with complete wording and instructions on the progression of the respondent through the form.

This style of layout is most commonly found in forms which are designed for self completion or where it is critical to the study that precise wording is used at every interview. It can lead to lengthy and complex questionnaires and is rarely found in agriculture or rural development surveys in developing countries.

- A listing of questions in a specific order, but without full or precise wording of the questions, or instructions for progression through the form.

This style of layout is frequently used in developing countries where the form will be completed by a trained enumerator. The form is normally accompanied by a detailed reference manual in which questions are specified in full and examples given. Careful training is necessary to ensure that enumerators follow the guidelines when they interview respondents. A major disadvantage of the approach is the problem of ensuring consistency between enumerators, especially where it is necessary to cross-question respondents to ensure answers have not been forgotten. The design of response categories can be used to help overcome this problem but enumerators never ask questions in identical phrasing, because questions are asked as part of a conversation, not as a list of topics.

- A tabular row and column format in which spaces are indicated for response (usually in coded form) without any specification of questions. Question order is indicated by the sequence of response categories.

This style is designed specifically to accommodate the needs of data processing. Great demands are placed on a reference manual and comprehensive training, and experienced enumerators are essential if it is to produce satisfactory results.

- A checklist of topics, indicating key facts to be covered, but with answers recorded either in an unstructured way in a field notebook, or in a simplified row/column table.

This last example is not, strictly, for a questionnaire, but for use in an informal study, such as a reconnaissance survey, in preparation for further investigation. It would be used by scientists and other experienced workers and never by enumerators. The content might indicate a maximum coverage of topics from which the interviewer would use his or her discretion to direct the interview with any single respondent.

Question phrasing

The information required should be well and clearly defined at each stage at which a question is posed: initial definition and explanation in the survey manual; text in the questionnaire; precise units for physical measurement; and verbal phraseology by the enumerator. Enumerator training must cover the precise meaning of questions to ensure that enumerators share a common understanding before they start interviews. At each stage the question should have:

- a clear meaning

- the same meaning to every person asked
- an answer which the respondent knows
- an answer which can be given clearly and unambiguously by the respondent

In Chapter 4 we looked at general principles of interviews and these are directly related to setting questions. In particular, it is important that the respondent can be reasonably expected to know the answer being sought – which is why it is always preferable to question doers, rather than observers of activities or events. Equally, if the interview involves a memory recall of a past event, there must be a reason for the respondent to recall the event in the detail being requested. This was illustrated with specific reference to labour data in Chapter 6.

Common problems which arise with question phrasing concern both syntax and content and are considered under the following headings:

- open or closed statements
- leading questions
- multiple questions
- ambiguous questions
- probing questions
- jargon and language
- sensitive topics

Open or closed statements

The simplest alternative styles of question are open or closed statements. An open question permits an open-ended response without any limitation on the range or complexity of the answer. For example:

'Which crops do you grow?'

A valid response here could include information for the current season, or for every season in recent memory, and could equally range from a simple list to a description of crop mixtures, planting times, and relative success or failure. The respondent is given an unconstrained opportunity, but this may not be what the survey wants. More often, the question would be modified to focus the answer on a specific time or place.

'Which crops are you growing this season?'
'Which crops did you grow last season?'
'Which crops did you grow on this plot last season?'

In these examples it is important to define precisely the modifying specification. What exactly is meant by season, especially when multiple or relay cropping is practised? And what are the boundaries of the plot of land?

More modification of the wording can result in a closed question. A completely closed question restricts the answer to Yes or No.

> 'Did you weed this plot last week?'

The response is not informative by itself, but a closed question can be used as an entry to a set of questions which apply only to certain respondents.

> 'Are you growing any sorghum this season?'

If No go to the next question
If Yes

> 'How many acres have you planted?'
> 'Which varieties are you growing?'
> 'Have you applied any chemical fertiliser?'

If Yes

> 'Which type(s) of fertiliser have you used?'
> 'List the quantity of each type of fertiliser that you applied to your sorghum'

The question about growing sorghum restricts the following questions to sorghum growers only. Similarly, questions about type and quantity of fertiliser are asked only to those farmers who used chemical fertiliser.

Open questions are valuable for pilot or exploratory surveys, when the range of possible response is unknown, and for informal enquiries which permit the interviewer to make notes in as much detail as required. They are not so useful for surveys undertaken by enumerators. Closed questions produce little information, but they are a useful way of separating out sections of the interview which do not apply to all respondents. They also permit complex questions with multiple clauses to be broken down into concise subjects.

Leading questions

The presentation of questions should be neutral. The form of the question should not indicate a preferred or 'correct' answer. The question

'You buy fertiliser don't you?'

contains a leading suggestion that you should buy fertiliser and that you're wrong if you fail to do so. The question

'Do you buy fertiliser?'

does not try to lead the respondent to a particular answer and contains less implied criticism of a negative answer. The question

'Do you buy the fertiliser recommended by the extension worker?'

implies once again that you should do so. Better to ask about the type of fertiliser purchased and then compare this with the extension recommendations.

'Do you buy fertiliser?'

Multiple questions

Ask only one question at a time. The question

'Do you have a tractor or a plough?'

does not permit a reasonable response from someone who has one item but not the other. Any question containing 'or' or 'and' should be checked to ensure that it does not ask two questions in one. Similarly, questions with these terms should be checked to make sure that they are not too complicated.

'Do you go to the nearest market on your bicycle more than once a week and take your wife and more than 5 kilos of produce to sell?'

This question is far too long. It should be broken down into a set of shorter questions in which a negative response to each part means that subsequent parts need not be asked. For example, if the first part of the above question asked 'do you go to the nearest market more than once per week?' and the answer was 'No' then the remaining elements may not be relevant. And not least, who or what is for sale?

Ambiguous questions

Ambiguous questions must be avoided in a survey since their interpretation will depend on the individual respondents. The question

'Do you usually sell your agricultural products in market X'

depends upon the interpretation of the term 'usually'. Even the apparently straightforward question

'How do you get to your local market'

may give answers which describe the means (bus, bicycle, etc) or the route. The question

'Do you use fertiliser?'

may also be ambiguous. Does the question refer only to artificial fertiliser or does it include animal dung? Some questions may be ambiguous only in particular circumstances and as long as the interviewer is well trained then he or she should be able to correct any misconceptions.

Probing questions

Good enumeration requires the ability to consider the response given to earlier questions and rephrase the subject in order either to cross-check the response or to elicit more detail. These questions can rarely be specified on the survey form, but should be explained in the manual and practised during training. The idea behind probing is for the enumerator to assess the validity of the response being given, either in terms of a general understanding of the survey topic, or with reference to prvious answers. For example, if it is known that families in a community keep small livestock, but are reluctant to admit the truth about numbers and types, a well-trained enumerator will not accept the first reply given without looking for confirmatory evidence. This may be visual such as goats and chickens in the compound, or the sight of a young boy tending cattle, or it may be based on other survey answers, such as a previous question about sales of animals at a local market. Sometimes the questionnaire can be designed to approach the same information from different aspects. Thus when the members of a family are listed, the opportunity can be taken to ask about income-related activities. Later in the interview, when income is being discussed, the response to the earlier question can be used to nudge a flagging memory.

Probing is not easy. A delicate balance has to be struck between persistence and rudeness. Very often the respondent does not want to tell the truth. In some cultures it is socially acceptable to tell lies to close friends, never mind strangers. The enumerator working on a repeated visit survey has to maintain a working relationship with the respondent and cannot permit the need to resolve minor contradictions on a few questions to disrupt the relationship. In some cases unbelievable data have to be accepted, and it is helpful if some method is agreed for the enumerator to draw attention to this on the form.

Jargon and language

The wording of the question must be appropriate to the respondent. Questions should avoid the use of technical terms which the respondent may not understand. Respondents may fail to ask for clarification if they feel that this will make them appear ignorant and may give an answer even if they do not fully understand the question. Also avoid treating the respondent as if he or she was stupid. Thus rather than asking about 'inorganic' fertiliser, the question should specify types or brand names or colloquial terms with which the respondent will be familiar. Similarly, questions should use as few and as short words as possible. Rather than ask about a 'remuneration level' ask about 'pay'. Use terms which the respondent will understand and which will not cause offence. Terms such as 'peasant', 'tribe' or 'witchdoctor' may cause offence.

Frequently, it will be necessary to make use of words and phrases from local languages which refer to specific events or customs. An example is the different arrangements for hiring draught oxen in parts of Ethiopia.

> *Mekanajo* refers to an agreement whereby two farmers with one animal each agree to pair their animals for work on alternate days.
> *Kollo* is a rental agreement for one ox, with payment in grain at a fixed quantity.
> *Domegna, yemoyategna and balegn* agreements exchange labour for the hire of oxen.
> *Yekul* is the system used by the poorest of farmers, who ask another farmer to plough their land and pay a proportion of their output in return.

Isolated words such as these can be incorporated in a questionnaire designed in English, but they must be defined in the reference manual. Most importantly, the surveyor should never assume that terms in local languages will all be recognised by enumerators who speak the language, or that everyone would agree on the same interpretation. Discuss translation during training and make sure all enumerators reach a consensus before fieldwork starts.

Sometimes the need for precise phraseology means that local languages and scripts have to be used even where English is the lingua franca. The alternatives are either to write the form completely in the local language, or to use a system of parallel translation of key phrases from the working language. Examples 7.1 and 7.2 illustrate two different styles of translation. In the first example, from Pakistan, Urdu phrases are written alongside the English. In the second example, from Papua New Guinea, the form is written in Tok Pisin, with English sub headings.

EXAMPLE 7.1

33 If you are in need of money, in what order would you approach the following:

اگر کبھی آپ کو روپیوں کی ضرورت ہو تو
آپ مندرجہ ذیل کس طریقے کو اپنائینگے
ترتیب وار بتائیے

(a) Relatives ☐
(b) Friends دوستوں کے ذریعے ☐
(c) Bank or Co-operative Society ☐
(d) Any other: (state) ________________

Code (33a–d)	
First choice	**1**
Second choice	**2**
Third choice	**3**

EXAMPLE 7.2

Aid posts .... Dei .... Mun .... 19......

Out-patients:	**Namba sikman kamap nambawan taim.**	....
	Namba bilong olgeta sikman bilong dispela mun	....
In-patients:	**Namba bilong sikman slip long wod**	....
Births:	**Hamas nupela pikinini kamap long ples bilong yu**	....
Deaths:	**Hamas mama i dai long karim pikinini**	....
	Hamas pikinini mama karim na dai long nambawan dei	....
	Hamas i kisim marasin na i dai yet	....
	Hamas i dai long Aid-Post	....

Sensitive Topics

As we have noted earlier, in many societies topics which are considered sensitive may appear to outsiders to be unremarkable. Because they cannot easily be predetermined, care has to be taken to consult people who are knowledgable about the culture.

Sensitive questions are prone to normative answers, answers which confirm that the respondent acts within the social rules of society even if that particular individual sometimes acts outside these rules. In a society which generally condemns drunkenness, questions about drunkenness might generate denial even if drunkenness sometimes does occur. Under these circumstances it may be useful to word the question so that there is some assumption that the activity does take place. Thus rather than ask 'do you ever get drunk?' we might ask 'how often do you get drunk?' The assumption in the question that you might sometimes get drunk may ease the guilt of the respondent and generate a more truthful answer.

Occasionally though, careful wording will not be enough. If the survey demands treatment of a topic for which honest answers cannot reasonably be expected one option is to use a form of randomised response technique. This approach is outside the main thrust of this book, which concentrates on factual data, but the principle is straightforward to apply and some readers may find it useful.

The technique is designed to disguise the response given to the interviewer so that the respondent is able to give an honest answer without the interviewer knowing. Answers are based on the fall of a coin (or other binary random choice). According to the random selection the answer given is either the sensitive response or the truth, but the interviewer does not know which (see Example 7.3).

EXAMPLE 7.3

Consider a study concerned with the distribution of farm inputs through a government controlled supply company with a fixed quota distribution. There is much evidence that store managers operate a black-market system of sales in excess of quota to traders. The whole study includes interviews with farmers, market samples, and a case study group of traders. There is great pressure to find out whether the supplies are originating from a few store managers or if all stores are involved. But if a question is posed asking if a store manager sold any

inputs above trading quotas it is expected that most managers would reply 'no', because they could be prosecuted if the government found out.

Instead, a randomised response technique is used. Each store manager is asked to spin a coin out of sight of the interviewer. He is told that if the coin shows heads he should reply 'yes' that he sold over quota, irrespective of whether he did or not. If the coin shows tails he should tell the truth and answer yes or no.

At analysis it can be predicted (unless the coin was biased) that 50 per cent of replies will be 'yes' because of the fall of the coin. Positive answers above 50 percent indicate the likely true response from the sample. Thus if the results from the store managers gave a positive response of 80%, we can conclude that 50% is due to the random effect of the coin, and the further 30% represents the true response. Because this effectively applies just to half the sample we would report that the indication was that 60% of store managers sold above quota and therefore the illicit trade is widespread.

The technique is simple in concept. To be successful a number of conditions should exist. First, the respondents must in principle be sufficiently in agreement with the aims of the study to give honest answers. Second, the sample of respondents should be large enough so that the probability of the fall of the coin is dependable. The limitations are that information cannot be associated with any other findings from individual respondents, because the truth of an individual's response is never known; and that the question needs to be posed as a simple closed response question. A multiple choice answer would need to be broken down into several statements.

Summary

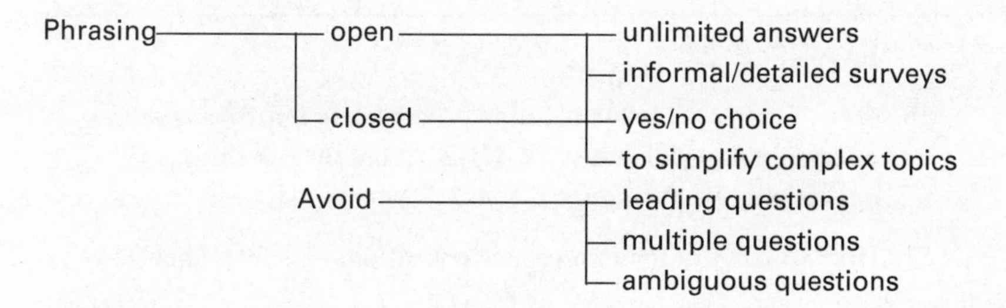

Probe for inconsistencies/unbelievable answers but, balance persistence against rudeness. Some information is unobtainable.

Sensitive topics — related to people's culture
— not always obvious

7.3 Answers

Response layout

Response options are classified into 'open' or 'closed'. Open implies that the respondent is permitted to answer in any form and at any length, to the question asked. Closed response implies that the respondents' answers are restricted in some way to a limited range of alternatives. Response layout is mostly concerned with the way in which closed categories are handled.

Open response

Open response categories are most often associated with exploratory or informal surveys, in which the surveyor does not know the likely response from the units of study. The design requires that sufficient space is left after each question for an answer. There is a danger that this would lead to excessively long forms so provision is often made to transfer information to the reverse of pages, or to a separate notebook. Open categories are often used for individual questions, perhaps seeking attitudinal or preference information, but rarely for a whole questionnaire. Particular problems arise in the analysis of open categories if the analysis requires response to be summarised or converted to a proportion, because each individual answer can differ markedly. They are more likely to be found in a pilot or exploratory study to define the likely response for categories in the main study. Example 7.4 shows the layout for some exploratory questions about institutional training.

EXAMPLE 7.4

16. One of the reasons for training officers is to improve the operational performance of AIDB. What do you think operational performance means?

17. How would you try to measure operational performance?

18. Do you think the training you received has made any change to the operational performance of AIDB?

19. Explain the changes which you think have taken place.

20. Training courses have benefits for individuals and for the organisation. Do you think the course you attended brought any benefits for AIDB? If YES – explain.

21. What changes would you make to future training to bring greater benefits to AIDB?

The advantages of open-ended response are:

- they may reveal responses which the questioner had not considered (and which, therefore, would not have appeared as closed-ended categories)
- they allow the respondent to indicate his or her exact rather than general response
- they may be used when there are too many response categories to list on a questionnaire
- they are useful when the questions are too complex to reduce to a few standard responses, and
- they allow the respondent to express him or herself in his or her own words

The disadvantages of open-ended response are:

- that much irrelevant information is collected
- the answers are not standardised and are therefore difficult to compare,
- coding is difficult
- they require a higher level of skill on the part of the respondent and enumerator

- the responses are often too general to be useful, and
- more time is needed for completion and the forms are often bulky

Closed response

Closed response questions require the enumerator or respondent to select an answer from a restricted list. The choice can be made by making a mark alongside a category; by entering a numeric value; or by selecting a code from a code list, as shown in Example 7.5.

EXAMPLE 7.5

A.	**Tick the appropriate blank**	**Sex: Male....**	**Female.....**
B.	**Tick the appropriate box**	**Sex: Male ☐**	**Female ☐**
C.	**Circle the appropriate code**	**Sex: Male 1**	**Female 2**
D.	**Enter the appropriate code**	**Sex: Male 1**	**Female 2 ☐**

Where there are many categories it is neatest to use a vertical list. With the exception of absolute categories, such as sex, every list should contain a provision for: Don't Know; Other (ie. not one of the answers specified); and Not Applicable. Example 7.6 illustrates some simple coding from a groundnut questionnaire. Every respondent (in this case the operator of a plot or garden) must answer every question, and provision is made for this with the categories 'Other' and 'Unknown' for question 4, and 'Insignificant' for question 22. The coded value will be transferred to the data processing medium, but in these examples the measurement is based on an ordinal system, described in Chapter 10. The codes have no numeric significance in terms of absolute or relative magnitude. Instead of the values 0–6, any other range, such as 10–16, could have been used.

The advantages of closed response categories are that:

- the answers are standardised and easy to compare
- the answers are easier to code and analyse
- the question meaning is often made more clear by the response categories, and
- the answers are relatively complete as long as all relevant categories are specified

EXAMPLE 7.6

5. What type of groundnuts are grown in the yield subplot?

Kalisere		**0**	
Mwitunde		**1**	
Chalimbana		**2**	
Malimba		**3**	
Mani Pintar		**4**	**(22)**
other (specify)		**5**	
mixture (specify)		**6**	
unknown		**7**	

22. Diseases in the yield subplot:

insignificant		**0**	
Rosette disease light		**1**	
leaf spot disease light		**2**	
combination (1–2) light		**3**	**(71)**
Rosette disease heavy		**4**	
leaf spot disease heavy		**5**	
combination (4–5) heavy		**6**	

In addition, respondents may find it easier to choose an answer from the categories rather than think up an entirely original one and they may be more willing to answer sensitive questions (such as livestock ownership or income) if the answer is in the form of a range category (i.e.

20–39 head, 40–59 head, etc.) rather than the absolute answer required from an open-ended equivalent question. There is a danger, if response categories are prompted, that the choice of a category is treated as a quiz for which there are preferred or correct answers. If later respondents overhear an interview this danger can arise, resulting in a succession of identical, 'acceptable' responses.

The disadvantages of a closed response are that:

- the respondents can guess at answers when they don't know, since they have the categories to guide them
- the appropriate category may be missing from the schedule
- failure to understand the question is less easily detected than with an open-ended question
- a poorly planned list may act as a constraint to correct answers not catered for
- too few categories may fail to differentiate between important groups, and
- enumerator error (placing the tick in the wrong box by accident) will be more common

Summary

Open response
- informal surveys
- very detailed
- hard to analyse

Closed response
- skill in setting categories
- can limit range of answers
- suits computer processing

7.4 Coding

Coded categories

Closed response categories are adequate as long as the number of alternative answers is limited. The lists in Example 7.6 occupy a reasonable amount of paper, and clearly many questions would result in a long document. In this case, choice of a code from a code list may be a better alternative. Example 7.7 shows one system for concentrating information about a large number of individual family members on one page. The problem facing the designer was how to accommodate a number of alternative reasons for arrival, departure or sickness, without needing a separate question for every family member. The solution

EXAMPLE 7.7

Changes in pot composition

Has anyone arrived or left or been sick in this pot since the last interview? Yes/No   If Yes specify below:

Code 800 reasons for arrivals

800	**Birth of child into family**
801	**Adoption/care of child**
802	**Marriage bringing new wife into family**
803	**Return from hospitalisation**
804	**Return from school or college**
805	**Relative joining family to work**
806	**Visitor**
807	**Other – specify**

Arrivals			
Names	Sex M/F	Age	Reason code 800

Code 820 reasons for departures

820 **Death of someone in the family**
821 **Divorce of wife from family**
822 **Hospitalisation**
823 **Marriage taking someone from the pot**
824 **Child leaving for school/college etc**
825 **Child leaving for another family**
826 **Relative leaving family to form his own pot**
827 **Other – specify**

Departure			
Names	Sex M/F	Age	Reason code 820

Code 840 reasons for sickness

840 **Injury on farm**
841 **Injury when not on farm**
842 **Fever/Malaria**
843 **Coughs/Chest pain**
844 **Diarrhoea/Stomach pains**
845 **Pregnancy/Delivery**
846 **Eye infection/Blindness**
847 **Other – specify**

Sickness				
Names	Sex M/F	Age	Reason code 840	No. days sick

adopted was to print code lists of the reason categories and enter the correct code next to each person's name.

Coding systems

Simple closed response categories make use of codes to identify each answer, but the meaning of the code is dependent on knowing the specific question. Using the system of Example 7.6, codes 0 to 5 would be repeated on many of the questions. With surveys that deal with repeated activities such as production, consumption, labour, income and expenditure, the information being collected often refers to a common set of items: crops, labour activities, foodstuffs, and so on. A standard code list would permit a unique set of code values for these items, and is often adopted in conjunction with questionnaires with extensive row/column formats. Example 7.8 shows a recording system for which the enumerator has to make use of a mixed set of codes. Some are universal and apply throughout the survey (week codes and crop codes). Others relate to specific topics on that page (land preparation type and method, weeding method). The standard codes means that the data are decipherable without necessarily knowing the question, and that enumerators will become familiar with the code values.

It is helpful if the master code list is designed for later addition of codes not included at first, and if codes are assigned in a logical sequence and grouping. Example 7.8 blocks crops first by traditional and improved, and then by grains, legumes, roots, etc. There is space in each category for the later addition of new crops or varieties.

Verbal answers have to be given numeric codes for ease of analysis. Manual processing can cope with verbal answers but computers are designed around numeric systems. The transfer of data from the verbal form (oral or written) to numeric is another data transfer and as such it is a potential source of error. The codes ticked in Example 7.6 can be checked during processing that they are in the correct range of valid codes. But that check will not confirm whether or not the enumerator ticked the correct category. The more extensive the system of codes the greater the chance that the wrong code is entered, or that a mis-written code is considered valid.

The form in Example 7.9 was designed for direct coding by the enumerator during the field interview. There is no possibility of confirming the correctness of the codes at a later stage. This design may be satisfactory in certain limited situations where the enumerators are experienced, or the subject matter is well understood, but it is not recommended for typical agricultural data collection. A good design has provision for a written response plus a coded value. Example 7.10 illustrates this dual system, which can also be seen in Example 7.8. The enumerator records the farmer's response in words during the interview and coding is done later.

Coding can be carried out by the enumerator but it is a convenient exercise for a supervisor. It ensures that the immediate supervisor has regular contact with the enumerator, and that every form is inspected. Under some surveys it may be preferable to defer coding until forms are returned to the office. The disadvantage here is that a longer period of time might elapse between inteview and coding, with less chance of the enumerator remebering any aspect of the interview that is queried. Also, if the forms are coded in the field errors can be corrected by a repeat interview swiftly and less costly than for a return from the office.

Summary

Coding
- Best if words and codes together on form
- Block code lists for later addition
- Link coding to supervision/checking

7.5 Layout

Form design

The final layout of the form must take into account the number of topics to be covered, the preferences of the enumerators and the circumstances in which the interviews will take place. With experience, surveyors tend to develop strong design and layout preferences and it is these which will govern the format chosen. The amount of effort given to layout and production should reflect the purpose of the questionnaire. A short one-off enquiry may be adequately produced by stencil on low-quality paper. A form to be re-used over several years, and/or by several organisations would justify printing on better quality paper, and possibly an input from a graphic designer on layout and presentation.

An important consideration since the spread of microcomputers is the availability of data management software which includes form design facilities. If a survey is to be analysed by computer it may be helpful for the data entry operators if the data entry screens are designed to represent the survey form as closely as possible. If this

EXAMPLE 7.8

Plot cultivation

Type of land preparation	Tick One
Flat	
Ridges	
Mounds or heaps	
Beds	

First season			
		Method	Date
Land Prep.	1		
	2		
Weeding or ridging-up	1		
	2		
	3		
	4		
	5		

To be completed by coder

Land preparation

Type ☐ Method ☐ Date ☐☐☐

Weeding

First Method ☐ Date ☐☐☐

Last Method ☐ Date ☐☐☐

Number ☐

Applying inputs

Fertiliser type	Qty	Unit	Date	Crop

Spraying of crop or crops in plot		Material	Date
	1		
	2		
	3		
	4		
	5		
	6		

To be completed by coder

Organic Manure

Quantity ☐☐☐☐ Date ☐☐☐

Inorganic fertiliser

Type	Kilos	Date	Crop
☐☐☐	☐☐☐	☐☐☐	☐☐☐
☐☐☐	☐☐☐	☐☐☐	☐☐☐
☐☐☐	☐☐☐	☐☐☐	☐☐☐
☐☐☐	☐☐☐	☐☐☐	☐☐☐

Spraying

Material	Date First	Num.
☐☐☐	☐☐☐	☐

EXAMPLE 7.8 continued

Week codes 1983/84

Week beginning:

April	4	1983	366
”	11	”	367
”	18	”	368
”	25	”	369
May	2	”	370
”	9	”	371
”	16	”	372
”	23	”	373
”	30	”	374
June	6	”	375
”	13	”	376
”	20	”	377
”	27	”	378
July	4	”	379
”	11	”	380
”	18	”	381
”	25	”	382
August	1	”	383
”	6	”	384
”	15	”	385
”	22	”	386
”	29	”	387
Sept	5	”	388
”	12	”	389
”	19	”	390
”	26	”	391
October	3	”	392
”	12	”	393
”	17	”	394
”	24	”	395
”	31	”	396
November	7	”	397
”	14	”	398
”	21	”	399
”	23	”	400
December	5	”	401
”	12	”	402
”	19	”	403
”	26	”	404
January	2	1984	405
”	9	”	406
”	16	”	407
”	23	”	408
”	30	”	409
February	6	”	410
”	12	”	411
”	20	”	412
”	27	”	413
March	5	”	414
”	12	”	415
”	19	”	416
”	26	”	417

Crop codes

	Trad.	*Impr*
Grains		
Acha	100	200
Guinea corn	101	201
Maize	102	212
Millet (early)	103	203
Millet (late)	104	204
Rice	105	205
Legumes		
Bambara nuts	110	210
Beans	111	211
Cowpeas	112	212
Pigeon peas	113	213
Groundnuts	114	214
Soyabeans	115	215
Roots		
Cassava	120	220
Cocoyam	121	221
Irish potatoes	122	222
Sweet potato	123	223
Yam (unspec)	126	226
Water yam	127	227
Yam (white)	128	228
Yam (yellow)	129	229
Fruit and vegetables		
Bitter tomatoes	130	230
Garden egg	131	231
Melons	132	232
Okra	133	233
Onions	134	234
Peppers	135	235
Pumpkin/ squash	136	236
Pineapple	137	237
Tomatoes	138	238
Vegetables – leafy	139	239
Vegetables – other	140	240
Sweet peppers	141	241
Other annuals		
Benniseed	150	250
Calabash	151	251
Castor oil	152	252
Cotton	153	253
Spices	154	254
Stalks/grass	155	255
Sugar cane	156	256
Tobacco	157	257
Hemp/rama/ kenaf	159	258
Tiger nut	159	259
Whole plot	199	

Land preparation type	
Ridges	1
Mounds or heaps	4
Flat	7
Other	9
Land preparation method	
None	1
Hand	2
Oxen	3
Tractor	4
Oxen/hand	5
Tractor/hand	6
Tractor/oxen	7
Tractor/oxen/hand	8
Other	9
Weeding method	
Hand	2
Oxen	3
Tractor	4
Chemical (herbicide)	9
Inorganic fertilisers	
Nitrates	
Sulfa	363
C.A.N.	364
Urea	363
Phosphates	
Supa	366
Boronated supa	367
Compound	368
Others (e.g. muriate of potash)	369
Spraying material	
Fungicide	380
Herbicide	381
Insecticide	382

approach is used the data entry screens should be laid out at the same time that the form is designed, to ensure compatibility. This point is explored further in Chapter 9.

Numeric response

The examples in this chapter have concentrated on written and coded answers. But for many subjects where a measurement is made a direct numeric answer is recorded. The consideration for layout is whether or not further computations have to be made after the data are collected, or whether the response will be used directly for analysis. A typical example concerns the counting of plants and planting locations (stands) from a subplot laid in a farmer's field. Example 7.11 is from an agronomic survey. Information is collected for up to four crops. The top of the page records general planting data. The middle section has 20 rows to record the number of stands and plants row by row in the sub-plot. The count of stands and plants are summed and the answers written at the foot of the columns. The bottom section is completed when codes are entered for data processing.

Measurements of crop output and the application of inputs may require more calculations of application rate and shelling, threshing or moisture percentage. Example 7.12 shows a layout for recording the application of sulphur dust and calculating the shelling percentage for groundnuts. Both examples, 7.11 and 7.12, require every stage in the measurement or calculation to be recorded on the form. It is better if working in note books or scrap paper, which are difficult to recheck after the form is submitted, is discouraged.

In surveys which involve multiple visits, information may be collected piecemeal as events take place. Good form layout can help to ensure that events which occur at unpredictable dates are not overlooked. A simple solution is to group events sequentially in a table, so that inspection of the form shows which items have been recorded. Example 7.13 applies this to crop production.

Information recorded about each crop and plot is stored sequentially so that missing items are instantly visible.

Another guide to enumerators is to draw arrows leading the flow of topics according to the response given to earlier questions. This style is most common on forms for self-completion, but it can equally be used with enumerators.

EXAMPLE 7.9

Strictly confidential

Deadline survey 1978–1979 (Crop Year April 1978–March 1979) A.P.M.E.P.U., P.M.B. 2178 Kaduna

Project ________ **Unguwa** ________ [1] [| |] **H/H Head** ________ [4] [|] **Field/plot no.** [6] [| | |]

Card no. [10] [1 | 0 | 1]

Field/Plot Area (ha) [13] [| | |] **Distance from Unguwa** [17] [|] **Distance from motor road (km)** [19] [|]

Crop planting order and place [21] [| |] (1) [24] [| |] (2) [27] [| |] (3) [30] [| |] (4) [33] [| |] (5) [36] [| |] (6)

1

Registration | | | | | | | | | 2 |

Card No.		Week No.		Act			Seed						Taki and Fert.							Chemicals							E	Crop Yield						Residue						
				T		M	C		Q		U		C		T	Q		U		C		T	Q		U		X	C		Q		U		C		T	Q		U	
1	2	3	4	5	6	7	8	9	2	1	2	3	4	5	6	7	8	9	3	1	2	3	4	5	6	7	8	9	4	1	2	3	4	5	6	7	8	9	5	1
1									0										0										0										0	
0	1																																							
0	2																																							
0	3																																							
0	4																																							
0	5																																							
0	6																																							
0	7																																							
0	8																																							
0	9																																							
1	0																																							
1	1																																							
1	2																																							
1	3																																							
1	4																																							
1	5																																							

Not to be punched

Other information

Week No.	Rain fall

EXAMPLE 7.10

Arrivals and departures

To be completed by enumerator

Arrivals

Week	Name of arrival	Age	Sex	Reason for arrival (see above list)	Normally available for work?

To be completed by senior enumerator

	1											12
Reg											0	2

Cd No.		Arr 1				A F	Arr 2				A F	Arr 3				A F	Arr 4				A F	Arr 5				A F
		Age		S	R	W	Age		S	R	W	Age		S	R	W	Age		S	R	W	Age		S	R	W
14	15	16	17	18	19	20	21	22	23	24	25	26	27	28	29	30	31	32	33	34	35	36	37	38	39	40
0	1																									
0	2																									

Departures

Week	Name of departure	Age	Sex	Reason for departure (see above list)	Normally available for work?

Reg	1											12
											0	3

Cd No.		Dep 1				A F	Dep 2				A F	Dep 3				A F	Dep 4				A F	Dep 5				A F
		Age		S	R	W	Age		S	R	W	Age		S	R	W	Age		S	R	W	Age		S	R	W
14	15	16	17	18	19	20	21	22	23	24	25	26	27	28	29	30	31	32	33	34	35	36	37	38	39	40
0	1																									
0	2																									

In this column (and in all subsequent columns marked WEEK) you record 1 for the first interview and 2 for the second interview.

EXAMPLE 7.11

FRADYS

Ward □□□ Household □□ Field □□ Plot □

Planting details

Crop				
Trad. or improved				
Planting or transplanting date				
Where planted				
Seed dressing				
Thinning date				
Triangle				

Stand count data

Date of count									
		St.	Pl.	St.	Pl.	St.	Pl.	St.	Pl.
Stands or plants by row	1								
	2								
	3								
	4								
	5								
	6								
	7								
Paces = ______	8								
÷ 2 = ______	9								
	10								
Random numbers	11								
	12								
1 ______	13								
2 ______	14								
3 ______	15								
4 ______	16								
	17								
	18								
	19								
	20								
Total stands									
Total plants									
Reasons for missing stand count									

To be completed by coder

Crop	Date plant	Wh. pl.	S-dr.	Date thin.	Triangle one stands	Triangle one plants	Triangle two stands	Triangle two plants
□□□	□□□	□	□	□□□	□□□	□□□	□□□	□□□

EXAMPLE 7.12

(spare box) ☐ (31)

11. How many times was sulphur dust applied on the y.s.p.? ☐☐ (32–33)

    (spare boxes) ☐☐☐ (34–36)

12. What weight of sulphur dust was applied on the whole garden/plot?

    no. of bags ☐☐☐ weight/bag ☐☐☐ kg ☐ lbs ☐ (tick) total weight ☐☐.☐ (37–39)

    (spare boxes) ☐☐☐☐☐☐☐☐☐☐☐ (40–50)

Harvest record

13. Total number of groundnut plants in the yield subplot ☐☐☐ (51–53)

    (spare boxes) ☐☐☐ (54–56)

14. Weight of groundnut pods at harvesting ☐☐.☐

15. Weight of groundnut pods after drying ☐☐.☐ (57–59)

16. Weight of kernels after shelling ☐☐.☐ (60–62)

L.L.D.P. only

weight of sample for shelling ☐☐.☐ lbs

weight of kernels from sample ☐☐.☐ lbs

shelling percentage ☐☐.☐ %

EXAMPLE 7.13

Crop	Field/ plot	Date plant	Date weed	No. of weeds	No. of stands	No. of plants	Fert. type	Fert. quant.	Chems used	Harvest weight	Plot area

EXAMPLE 7.14

Identification

serial number		card type	(1–2)
village		project	(3)
hh. head		stratum (a)/sample	(4–5)
sex	male / female	unit	(6–7)
garden no.	plot no.	village	(8–10)
operator		household	(11–12)
		oper's no./person no.	(13–14)
date		stratum (b)/ev.a.	(15–16)
enumerator		garden no. and plot no.	(17–18)

checking by:	date	initials	comments/errors:
supervisor			
ch/ver HQ			
punched			
verified			

Data processing

The coding system used must cater for the method of data processing. Manual tabulations can be made direct from verbal responses, but both manual and computer processing is easier from coded data. The code system should present the coded value in a way that is clear for the data entry clerks to use. This is helped if:

- coded response categories are right-adjusted in a column down the page, such as Example 7.6
- row column layouts do not include a mixture of information some of which is for processing and some not
- codes and measurements are written with an indication of how many digits or characters can be entered to a computer system. The boxes in Example 7.12 are a good illustration of this
- the location of the decimal point and the number of digits after the point are clearly indicated (as in Example 7.12)
- answers include some indication of their variable name to be used in processing, or their character position along the computer data record. Variable names can be seen in the bottom section of Example 7.11. The response categories in Example 7.6 have a bracketed number alongside which corresponds to the character position number

Identification and administration

Every form should start with a section for information about who the respondent is, their location or other sample information, the date, and the name of the interviewer. In situations where the respondent may not be known until the interview takes place it is helpful to indicate not just the name, but also the occupation of the respondent, such as household head, or eldest son, or cooperative treasurer, etc. Provision should also be made for identification of the type of survey as a code for computer processing, and for the stages of coding, checking and computer or manual data entry. Example 7.14 illustrates identification data.

Production

The form is the main working tool of the enumerator and efforts to ensure a clean, neat layout will be rewarded with better quality work. Short, simple surveys merit nothing more than photocopied or duplicated pages. The more extensive the survey and questionnaire the more a high quality form is required. Printing and heavyweight, good quality paper, combine to produce a more durable product that is pleasant to use. A booklet format, of a size chosen to suit the enumerator's case or clipboard, often proves popular, and again is more durable than stapled sheets. As a final touch, colour coding of forms for different topics or dates is a great help to recognition both in the field and the office.

8

Field work

It is easy to think of field work as being the simple part in the process of obtaining, interpreting and using information. It is usually a relatively formalised and structured part of the whole operation, and can all too easily be regarded as the 'mechanical' part, in contrast with the more creative and intellectually demanding aspects of planning, and of analysis and interpretation. And so indeed it is – but it still requires the investigator's fullest attention, because it is the crucial, fundamental stage on which everything else depends. Faults and problems at other stages can be, at least to some extent, corrected and retrieved, but if incorrect information is gathered in the field, then nothing can compensate for that, and the whole operation is a failure.

Although there is very little in the way of theoretical argument or general principles which is of relevance when considering the organisation and management of field work, it is possible to identify some common aspects which often require consideration by the investigator. And although examples of specific solutions and procedures may not be appropriate outside their own particular circumstances, they may still suggest useful pointers and directions. This chapter therefore seeks to identify some of the common topics and problems in field work which often need to be addressed, and to give some examples of points of information and technique which may be of relevance.[1]

8.1 Organisation of field work

Field organisation for informal surveys

An informal, exploratory survey is relatively straightforward as regards the organisation of field workers. The fieldwork is normally done directly by high-level professional staff or researchers, who are very familiar with the subject, and highly motivated, so that training and supervision needs are minimal. These researchers are usually seconded from other work for a limited period in order to conduct the exploratory survey, so that long-term organisational issues do not arise. It may be necessary to arrange for the recruitment or secondment of interpreters in some cases.

A survey of this kind is usually conducted by a multi-disciplinary team. It is important to mix together workers of different disciplines, and to ensure that all the team members communicate freely with each other during the course of the fieldwork. The ways in which this was done in the original *sondeo* approach provide a good model (Hildebrand, 1981). In this work, researchers were teamed in pairs, each consisting of one natural scientist and one social scientist. The pairings were changed after each day, and discussions involving the whole team were held at the end of each day.

Field organisation for formal surveys

For formal surveys it is more common to have to use lower-level fieldworkers, and the quality of these is one of the most crucial factors in determining the quality of information. It affects the quality of the data at the point at which they are collected; errors at this point are the most difficult to detect, and the most difficult to correct or salvage at any later stage of the data collection process. Extra care and attention at the stage of recruitment of enumerators is likely to produce substantial dividends later.

We can distinguish three types of approach to the recruitment of field staff. We might recruit staff for a particular survey (i.e. for a limited period only); or establish a permanent body of field staff to conduct a continuous survey programme; or use an existing group of people, either from an established data collection organisation of some kind (a national or regional statistical office, a development agency or ministry, an extension service, or a commercial market research or polling company), or from a body with some interest in the subject of the survey (a university department, a voluntary association, a charity, or a school).

Each of these approaches has its advantages and disadvantages. In some situations, the choice of a particular approach is inevitable. For example, in the case of a census, either of population or of agriculture, the numbers of field staff needed are so large that every existing body or group of people which is at all appropriate may need to be utilised –

existing statistical organisations, schoolteachers and senior students, existing administrative staff of all kinds, and civil servants. It is sometimes convenient to use this type of approach, of using other organisations, for smaller-scale operations too. This is often the case where resources, either of funds or of suitable personnel, are very limited, and the subject of the survey in question is of particular relevance or interest to another organisation. Some examples of this might be an investigation of the incidence of cattle diseases and pests involving a veterinary college, or a survey of the take-up of new crop varieties by farmers involving an agricultural research station or an agricultural extension service. In this case, as well as increasing the available resources, there is the further advantage of using fieldworkers who should already have particular knowledge of the subject being studied. There is though a corresponding disadvantage, in that this particular knowledge may lead the enumerators to put questions in a leading fashion, or make prior assumptions about the information to be gathered, and thus bias the results towards their preconceptions. This is especially likely to occur if the enumerators' usual function involves the teaching or motivation of the population in the subject of the survey (as for example extension workers). The other major disadvantage of this approach is the difficulty of controlling a piece of work when operating through another organisation. Typically, this has to be done indirectly, with very little control of the organisational and supervisory processes. Ensuring and assessing data quality then becomes extremely difficult.

Full control of the selection, training and supervision of the field staff, and the implications of this as regards the quality of data collection, is the major advantage of having one's own body of enumerators. This advantage is significantly greater if it is possible to retain such a body of field staff on a permanent or reasonably long-term basis. It is then possible to set up an integrated, long-term programme of training and staff development, allowing for career development and so helping to produce good quality and well-motivated staff. A permanent field organisation has the advantage of a permanent system of administrative and logistic support in office facilities, transport, equipment, accommodation, channels of communication – and of contact with the local population for information and publicity. It is not of course practicable to set up a permanent organisation if it is simply an individual, 'one-off' survey which is being planned – but it is worth considering whether it may not be the first of a series of surveys. There might perhaps be other, related surveys being planned by another organisation, which might together form a logical programme of field work and would justify establishing a permanent body of enumerators.

A possible disadvantage of setting up a permanent fieldwork organisation is the potential conflicts with already established organisations. If such an organisation exists, and has an apparent overlap of function, it may well regard the establishment of a new organisation as threatening or disparaging its own work, and this may cause administrative or operational problems, particularly if both organisations are governmental or quasi-governmental bodies. In this situation tact is needed and it is sometimes helpful to stress a narrower or more specific focus of work on the part of the new body. It may well be worth considering setting up a formally-constituted coordinating and technical advisory committee on which already existing organisations can be represented.

It is difficult to maintain good organisation in the typical survey conditions of a dispersed enumeration force. Enumerators may have no fixed location, survey staff may only be temporarily employed and the organisation be somewhat *ad hoc*. It is therefore important to establish a very clear, explicit, complete and unambiguous structure of responsibilities, reporting and communication. It may prove helpful to draw up a detailed organisational chart, and to check from this that the distribution of responsibilities is appropriate: that each person is responsible to one and only one superior, that no-one has too many immediate subordinates for effective control, and that arrangements and communication facilities exist (or can be established) between each person and each of his or her immediate subordinates.

A duty statement, such as that given as Example 8.1, is often useful.

EXAMPLE 8.1

Duty statement

You are to work as a *Senior enumerator* in the Agronomic Survey.

You are responsible to the *Field officer*.

You are in charge of the five *Enumerators* working in the *Ekiti* Survey area.

You are responsible for the supervision of your team of enumerators, as described in the Agronomic Survey Supervisors' Manual, including:

General administration of the team
Organising the team's work
Checking work progress
Observing field work
Re-checking field work
Office checking of survey forms
Training and re-training of enumerators
Document control and record-keeping

Every month you are to report to the Field Officer in person, and to produce a Field Supervisor's Monthly Report.

You are to hold regular weekly meetings with your team of Enumerators, and also to make a random field visit to each of them at least once a week.

8.2 Recruitment of field workers

Characteristics of enumerators

Educational level

In selecting enumerators, various personal and acquired characteristics are desirable. One which is usually, and readily, defined is a certain minimum level of general education. This should relate to the level of education actually required by the enumerator in order to fulfil the survey functions – any higher may be counter-productive. While the level of intelligence and problem-solving capacity of a candidate may be related to his level of education, there are many circumstances which may weaken this relationship. There may be differential access to education for children from different areas of a country (rural children are often particularly disadvantaged, especially at secondary and tertiary levels), or for children from different socio-economic classes, or of different sexes. In some cases the type of education offered may even hinder the development of initiative and problem-solving techniques, for example if it is of a strong rote-learning orientation. Furthermore, candidates with a higher level of education may be unsuitable in some respects for work as enumerators. They may consider themselves to be over-qualified for the work, leading to a poor attitude both to the work and to the respondents, and a resultant lower level of data quality, or even data fabrication, and also to a higher level of staff turnover. In most situations, the best results are likely to be obtained by setting a minimum educational level according to the requirements of the work, and making further assessments of intelligence, skills and attitudes in the course of the selection procedure.

Formal skills

Certain formal skills are necessary for most enumeration work. Most obvious is a facility in simple *arithmetic*. This should include a knowledge of addition, subtraction, multiplication and division; a familiarity with the concepts of means, ratios, proportions and percentages and facility in calculating and manipulating them; and a knowledge of operations involving decimals and fractions. The need for a good level of *literacy* is equally clear – this is required for reading survey instruments and instructions, recording responses, and general administrative needs. In some countries, particularly where there are many vernacular languages, enumerators may need to be fluent in one or more vernaculars as well as the language used for administrative and technical purposes.

Other necessary skills are less likely to be the subject of formal qualifications, and must be assessed directly. One such skill is *measuring* – of using a measuring instrument to obtain an accurate and precise value of a characteristic. Of course, many of the instruments used in data collection are likely to be unfamiliar to prospective enumerators and will need special training in their use, but it is possible to assess basic skill in measurement (precision, interpolation, etc.) using simple and familiar instruments. A candidate who cannot use a ruler or a spring balance correctly is unlikely to be successful in using a compass or a moisture meter.

Another type of skill which normally needs to be taught to enumerators, and where their ability can be assessed if they do not already have experience, is in the *use of maps*. Enumeration work is likely to involve reading maps in order to identify and locate survey areas, and drawing sketch maps of enumeration areas, household locations or sample fields. An assessment of whether someone can deal with maps may be obtained by observing such simple tasks as locating a well-known place on a national or regional map.

Another skill which most prospective enumerators are unlikely to have been taught or to have practised previously is that of *interviewing*. This is therefore a skill which needs to be taught. However, there are certain characteristics which are important for good interviewing which cannot easily be taught, and these need to be assessed during enumerator selection. An interviewer should be fluent in speech, flexible in discussion, able to paraphrase, simplify and expand on a topic, and to adapt his or her level of approach to different respondents.

A characteristic which often proves invaluable for enumerators working in a context of agricultural development is a background of *agricultural knowledge* and experience. In the course of fieldwork it is often necessary to be able to identify agricultural commodities, operations or inputs from partial information or description (for example, to identify a type of fertiliser from a description of its physical appearance, or a chemical from an old container with partially obliterated label). A correct assessment of the plausibility and internal consistency of information given by respondents on agricultural matters generally requires a reasonable knowledge of agricultural systems and practices in the survey area. An additional, related point is that the attitude of farmers to an enumerator with little agricultural knowledge is likely to be one of derision, and grudging cooperation.

The above discussion does not imply that staff already working in an agriculture-related job are necessarily suitable as enumerators. In particular, a person in agriculturally-related work which involves teaching and advising, or selling, is likely to have developed attitudes about the superiority of modern methods which are not appropriate to enumeration work and which may have to be 'unlearnt' before a person with this type of background is suitable to be an enumerator.

One factor which is particularly likely to be associated with an agricultural, or at least a rural, background, is a willingness and ability to live and work in a rural environment. For many types of fieldwork an enumerator must either live for a substantial period in a rural community or travel within a rural area staying wherever is convenient, and it is necessary that he or she shall be able to adapt readily and willingly to the rural way of life and to the level of facilities available.

Personal characteristics

In view of the way of life a field worker leads, and the type of work done, certain traits of personality and character are particularly desirable in enumerators. Firstly, since the work is often performed alone, they must have a mature, honest and conscientious attitude towards their work and responsibilities. They must be able to motivate themselves to do the work without being under close supervision, and capable of organising their time effectively to be able to do so. They must be intelligent and self-confident enough to be able to deal with minor problems on their own initiative. At the same time, since there is always some element of team work in enumeration, an enumerator must be able to work effectively in a team with others, without a rigid hierarchy or division of labour being imposed. In the work itself, an enumerator must be capable of being meticulous and thorough in measuring or observing, and patient, courteous, flexible and perceptive in interviewing. Perhaps the most critical characteristic of all is the ability to learn.

It is often automatically assumed that enumerators must be male. In many societies there are social conventions and constraints that would make it difficult for female enumerators to live and work by themselves, or to travel and work as part of a mixed-sex team. Such considerations must depend on local conditions – but it is not uncommon to find that the received wisdom on such matters is lagging behind actual practice, and that the true situation is more flexible than is generally supposed. It is generally worth making an effort to assess thoroughly and without preconceptions whether it is possible to employ women, since selecting from a larger pool of candidates for employment will probably produce better staff. For surveys where information from women is being sought, it is useful to have at least some women enumerators. As regards both sex and age, it is preferable to select individual enumerators on the basis of the characteristics directly relevant to the work, rather than to exclude whole categories of people.

Similar general considerations apply when considering such characteristics as ethnic or tribal group, caste, or religion. It may often be most convenient to match the characteristics of the enumerators with those of the respondents, or of the population of the survey area in general. But it is worth making a careful assessment as to whether such restrictions of choice are really necessary.

Selection of enumerators

The selection procedure for enumerators needs to focus on assessing to the greatest extent possible, the characteristics discussed above. It is usually best to make a first selection by means of a written test. This is a relatively rapid and simple method of assessing some of the characteristics, and thereby selecting a smaller number of candidates for the subsequent, more time-consuming selection stage involving interviewing. The characteristics which can be assessed in a written test include:

- clear handwriting
- arithmetical skills
- language skills
- agricultural knowledge
- ability to extract information
- ability to approach problems

The last two, in particular, can only be very partially assessed by this approach, but some of the questions in the test can be set in such a way as to tackle these issues. It is often a good idea to start with a few simple, straightforward arithmetical questions, to settle and relax the candidates and check their arithmetic skills. Next there can be some questions which give quantitative information in the form of a narrative, and require certain extractions and calculations to be made. A simple example of this type of question is Example 8.2:

EXAMPLE 8.2

A farmer gives you some information on the amount of money she obtained from the sale of some of her crops. She says she sold some maize for $460, tomatoes for $125, and wheat for $560. For her cotton crop she got $620, and from the sale of onions she obtained $235.

(a) How much money did she obtain in total from the sale of crops?
(b) How much did she get in total for all her food crops?
(c) What percentage of the total money she obtained was from the sale of food crops?'

Before conducting the test it is preferable to set a minimum score for candidates to attain if they are to be interviewed, and to make this clear to them.

In the interview, the main points to be assessed are those relating to personality and character, and the way in which tasks are tackled, as discussed in the previous section. These must be assessed indirectly, by observing the candidate's manner, behaviour, and responses to questions during interviewing. This type of assessment can be combined with the direct assessment of other skills, such as measuring or map-reading. For example, a question asking a candidate to identify on a map their place of origin, or the measurement of the distance between two points, gives opportunities for assessing the candidate's way of approaching tasks in general. Example 8.3 gives another such question, based on arithmetic, but testing mainly how quickly points are observed and related to each other.

EXAMPLE 8.3

The following question can be used either entirely during the interview stage, or the earlier part can be included in the written test and followed up in the interview:

Consider the numbers 2,2,3,3,3,4,5.

What is the mean (average), median and mode of these numbers?

These concepts are normally covered in secondary school mathematics courses, and the calculations are very simple. The candidate's ability to apply known concepts is being tested. At the interview stage candidates are asked whether they notice anything special about the calculation of the mean of these numbers. The speed with which they notice that this value, 22/7, approximates 'pi', and the number of hints they need are useful points of assessment.

No matter how thorough and searching an interview, however (and it is seldom possible to spend a very long time over every individual interview), there will always be some chosen candidates who prove in practice to be unsuitable, or who find the work unsatisfactory. It is therefore necessary to make an allowance for wastage when making the initial selection, and to include a period of trial employment upon first recruitment of enumerators. During the trial employment, the new enumerator should work closely with an experienced enumerator (if the situation allows this), and should be closely supervised and continuously assessed by the field supervisor.

Senior enumerators and supervisors

Whenever possible, it is preferable that senior and supervisory field staff be recruited by internal promotion. This has the dual advantages that the candidates for promotion are well known, their character and performance have been assessed over a considerable period, and that the existence of prospects of promotion for enumerators makes it easier to recruit and retain good quality staff.

Supervisory staff, whether recruited by this or another method, and whether or not they have proven capacity as enumerators, will still in all cases need training in supervision. This should include training in the personnel management skills of motivating and leading enumerators, in the specific skills of supervising and checking field work, in organisation and record-keeping, and in the training of enumerators.

If the survey design allows, it is often found to be preferable for enumeration to be done in teams, each consisting of one supervisor and up to five or six enumerators. This makes the actual supervision and the organisational aspects such as record-keeping and document control easier, and also gives a faster response to enumeration problems. In addition, it allows the team to share the burden of any heavy concentrations of work there may be in the enumeration programme, and helps to maintain the motivation and morale of the enumerators.

The ratio of supervisors to enumerators should be very carefully considered. It should not be so low that not enough attention is paid by the supervisor to each individual enumerator, but equally it should not be so high that there are multiple layers of supervisors, all busily engaged in tripping over each others' feet. In practice, a ratio at each level of one supervisor to between five and seven of the next lowest level is effective.

8.3 Training of field workers

Enumerator training

There are several elements to enumerator training. The first and most obvious is that of familiarisation with the survey. This includes such general issues as the reasons why the survey is being held, its relevance to national and local development, the rationale of sampling (if relevant), the purpose of particular measurement techniques or topics included in the survey, and any place it may have within any larger programme of surveys.

Next the content of the survey needs to be considered. Each individual question or item of measurement needs to be discussed, and any issues of concepts, definitions, coverage, reference periods and inter-relationships between different questions all need to be fully explained. Very often, the explanation of survey questions to enumerators can act as a proxy for a pilot survey, testing whether the questions are easy to understand and answer, and appropriate to local agricultural conditions and farming practices. It is therefore important to discuss the questions at an early stage, and explore any problems of interpretation or measurement which might arise in the field.

The methods of data collection to be used in the survey must be included in the training. Particular attention needs to be paid to techniques of measurement involving special skills, such as the use of a compass to take bearings. Even familiar techniques, such as the use of a spring balance to weigh a quantity of produce, may not be as obvious to the enumerators, and all techniques which are to be used, no matter how simple, should be included in the training.

For measurement techniques or procedures which are particularly long or complex, or where it is difficult to arrange sufficient practical training, it is worth considering the use of videotapes, showing the procedures or techniques to be used. These can be used at any time of the year, do not require a field training site to be available, can be shown repeatedly as required, and can be shown by non-technical staff if necessary. Such training tapes can be commissioned from specialist film-making companies, but this is likely to be a relatively expensive option, and in-house production of training videotapes should at least be considered. The costs of this are now moderate, and it is worth investigating possible assistance for both training and funding from aid agencies, particularly those concerned with education and training.

Similar considerations apply to the teaching of arithmetical skills. It is unwise to assume that enumerators know, for example, how to add up a column of weights to give a correct total, or to add 180 degrees to a bearing to obtain a back bearing, or how to calculate from the farmer's reported quantities what proportion of a crop's total production has been sold. If an arithmetical technique is needed for conducting a survey it should be included in the training.

Specific training in interview techniques needs to be included for any survey where some of the data are to be gathered by interview. This includes general topics, including introducing oneself to the respondent and gaining their confidence, explaining the nature and purpose of the survey, explaining why the respondent has been selected and other people have not, and explaining the confidentiality of the information gathered. It should also include consideration of the way in which each individual question is best asked, with possible alternative phrasing and subsidiary questions if the respondent does not understand the question or has difficulty in recalling or expressing the answer. If different languages from that on the questionnaire are likely to be used during enumeration, then possible translations of questions, especially of any technical or special concepts, should be thoroughly discussed. Independent back-translation of the translated questions provides a useful check. It is also important to stress the need for a courteous, patient and respectful attitude in conducting interviews.

An important aspect of interview technique, but one which is likely to be inherent rather than one that can be taught, is the ability to assess the degree of completeness, consistency and plausibility of the responses. Some guidelines can be given about likely responses to particular questions, and some possible probing and checking questions for use with unlikely answers. Also, attention can be drawn to the relationships between different questions, and possible inconsistencies discussed (for example, too small an interval between the reported ages of parents and children). In some surveys some categories of information are particularly likely to be omitted – for example, if all household members are being listed, small children or young unmarried men may be missed out; and if all plots of cultivation are being recorded, those distant from the house or from the household's other plots may be left out. Possibilities of this kind can be identified and suitable probing questions framed.

After all the aspects discussed above have been thoroughly dealt with in the classroom it is very important to incorporate a large allowance of practical work in the training. First of all, role-playing practice in interviewing, and practice in any measurement techniques required by the survey, are necessary. After this, a further step in training, taking it even closer to a real enumeration, is to have the trainees conduct practice enumerations on real respondents, under the close observation and supervision of the trainers. It is very important that adequate time should be allowed in the training programme for these types of practical training – about half the total time is a good proportion.

It is essential to recognise that training does not stop when enumeration begins. Particularly in the early stages, a great deal of the supervision should be directed towards identifying problems and misconceptions, and providing training to overcome these. For a relatively long-term survey or programme, a regular programme of training is strongly advisable. This should cover problems observed during supervision, and refresher training on any topics or techniques which may be introduced at different stages or phases of the survey. For example, in a survey which gathers data on different agricultural operations throughout an agricultural season, refresher training on each stage (for example on land preparation, planting, fertiliser application, harvesting, etc.), should precede the actual gathering of data on that phase. A continuing programme of training is also a good vehicle for general morale-building of field staff, in showing that attention is paid to their work, in providing feed-back from earlier work, and in stressing the use and importance of the data gathered.

Enumerators' manual

An enumerators' manual has several different purposes, which are not easy to reconcile effectively. It is needed for use during the initial training period, and for reference purposes during subsequent field work and refresher training. It must therefore be fully comprehensive, so that it can be used to provide solutions in doubtful or complicated situations. At the same time, it must be easy to use, both to obtain a general grounding, and for referring to specific points. It will also need to be used by the enumerator for quick reference in the field during enumeration, and so needs to be of convenient size and robust construction, and simple to use for rapid reference purposes. There is therefore a good case for producing two enumerators' manuals – one, the manual proper, of a comprehensive and detailed nature for use during training, and for referring to to solve particular difficulties; and the other a condensed summary guide for rapid reference during field work. The latter level of information can sometimes be included on the survey form itself.

The full manual needs to cover all the topics discussed in the previous section. These include:

Context of the survey	– reasons for conducting it
	– relevance to development
	– rationale of sampling
	– purpose of particular questions
	– place within survey programme

Content of the survey	– relationships between the questions – for each question, the meaning concepts definitions coverage reference period
Techniques of data collection	– measurement techniques – interview techniques approaching respondents phrasing questions checking and probing

The use of the manual for reference purposes can be assisted by a well-considered layout. Firstly, the contents of the manual should be laid out in a clear and logical fashion, which should be fully described in a detailed table of contents at the beginning (as shown in Example 8.4).

EXAMPLE 8.4

Agronomic survey

Enumerators' manual

Contents

(continued)

Making your weekly plan
Keeping your diary
Dealing with problems

There are also, however, several points relating to the physical construction and layout of the manual which can be particularly helpful. Tabs or cut-outs can assist in identifying those sections of the manual dealing with particular topics, as can the use of different colours of paper for different sections, or of stiff dividers between sections. It is also sometimes helpful to place summaries of important sections, or examples, in conspicuous boxes or bold type. Computer word processing and graphics software include provision for a variety of different typefaces and sizes, together with designs and symbols which can be used to draw attention to the text. For a large-scale survey, it may be necessary to prepare similar manuals for supervisors and office staff.

8.4 Management of field work

Logistic organisation

A major element in the planning and organisation of a survey is ensuring that all necessary resources, equipment and materials are available at the correct place and the right time.

It is then necessary to plan and arrange for their procurement. In some cases this may involve borrowing facilities, such as school or training centre classrooms for training, or dormitories for accommodation. Sometimes it may be necessary to hire equipment or services, for example transport, or casual labour for assisting in field measurement or agricultural operations. And some materials and equipment, for example stationery, measuring tapes, or compasses, may require the purchase of supplies.

In the case of materials and equipment it is also important to plan and arrange their distribution to field staff. To avoid inextricable confusion, all such distribution should follow the channels defined by the organisational structure of the survey personnel, passing through each level of supervisory personnel in turn.

The most crucial type of resource needed for field operations is of course personnel, a subject which has been discussed in the previous sections. There are also some other types of resources, the need for which derives from the field staff. The most immediate and obvious of these is that of accommodation.

Accommodation requirements depend on the type of survey, and survey organisation, and the local conditions. If enumerators are moving from place to place during the course of the survey, then temporary accommodation needs to be arranged. If the enumerators are travelling within a relatively small area it may be possible for them to be accommodated at a fixed location, and to travel to the survey areas on a daily basis. If the survey requires the enumerator to remain in a fixed location or a small area for a long period, then more permanent accommodation is needed. Specific arrangements are very much dependent on local conditions.

The type of survey, the accommodation arrangements, and the local transport conditions, will all influence the transport requirements of the survey. Likewise, the implications of transport limitations must be considered when planning the workload of enumerators. For example, for enumerators in a fixed location, walking or travelling by bicycle may well be adequate. If teams of enumerators are travelling from place to place, then one suitable vehicle of adequate size is needed per team (and where necessary with four-wheel drive). For an individual enumerator covering a large area ill-supplied with public transport, or a supervisor covering several enumerators in scattered locations, a motor cycle may be the best solution.

Apart from transport, the largest items of equipment needed are likely to be measuring instruments of various kinds. The need for such items is survey-specific, but some measurement activities can occur quite frequently in survey programmes relating to agricultural development. One is the measurement of area, which will require compass, ranging poles, and measuring tape. Another is the weighing of produce, needing a spring balance (with a range and precision of scale appropriate to the items to be weighed) and a suitable container for the produce to be weighed in (usually either a special-purpose weighing sling, or a sack). If a liquid is to be measured (e.g. milk), then a light, unbreakable, translucent graduated container is suitable. In general, the identification of equipment needed must be done in very good time, as it is often rather specialised, and difficult, time-consuming or expensive to procure. For very specialised equipment for one-off purposes it may be possible to borrow from an institution such as an agricultural research station, university or agricultural college. Such institutions may also be helpful in doing measurements using their own facilities, for example in the analysis of soil or the moisture content of crops.

The most survey-specific equipment of all is the survey forms and questionnaires, and survey manuals and other documents. These must all be planned, designed, printed and distributed in good time. A document control system for the completed survey forms must also be established, and the necessary forms for this system, and any related summary data forms, must also be designed, produced and distributed. The question of stationery supplies is easy to forget or dismiss, but vital to remember.

For completing the questionnaires, the enumerators will require a supply of pens, pencils, erasers and pencil sharpeners. Writing boards or clip-boards will also be useful, since there is seldom a suitable writing surface available. Depending on local climatic conditions and the season, a waterproof cover for the survey documents may be needed. For many types of survey it is necessary to mark dwellings with a household identification number, which means that suitable materials for doing this, such as stickers of gummed paper, or chalk, need to be supplied to the enumerators. Finally, it is a good idea to supply the enumerator with a bag – large enough to carry all the survey materials, robust enough to cope with the likely field conditions, and of a suitable design for carrying on whatever mode of transport is used.

Public relations

It is impossible to conduct a survey without the active cooperation of the respondents. To gain this, the population affected by the survey must be given information about the survey, and in such a way as to make clear the desirability of cooperation by the respondents. It is also important to give people the opportunity to ask questions about the survey, and not simply be told about it.

The first thing which must be communicated is some understanding of what a survey is, since the concept itself is likely to be quite strange to most of the population. It is vital to stress the need for good information in order to achieve effective development planning. In this respect, it is often helpful to give examples of development projects which have been conducted in the survey area, or may be conducted in the future. But it is important not to give the impression that specific local development projects will appear as a consequence of the present survey exercise. (This kind of impression may be helpful in the very short term in gaining their cooperation, but is storing up considerable trouble for the future.) It is best to convey the need for economic development to be based on facts.

It may be necessary to point out also that there is no assessment for taxation of any kind involved, and to stress the survey's connection with the sponsoring authority.

In the case of a sample survey, suspicion and lack of cooperation can arise because the respondent sees that he has been selected for this troublesome exercise, whereas his neighbours have not. Why is he being picked upon? It is relatively easy to explain that, since it is not possible to enumerate everyone, some have been selected to represent all. An analogy with tasting a spoonful of a pot of food in order to check seasoning of the whole is often effective in making this clear. Also not easy to understand is why the selection happened upon one particular respondent, and why it cannot simply be transferred to his neighbour. An explanation that the selection has been done in a scientific fashion, to ensure that all types of person, household or farm are correctly represented, may be effective.

There is a wide variety of means of conveying these messages, and all possible channels should be used. It is important to inform and involve the traditional community hierarchy, and local government and administrative personnel. Their cooperation is essential in giving local credibility to the survey work. Often they can also be helpful in providing facilities or giving information on local conditions, and may even be obstructive if not kept fully informed.

Channels of communication through local organisations are also useful for publicity purposes. Local social and community groups, farmers' organisations, and cooperatives may all be appropriate. Political, religious or ethnic organisations may in some cases be useful, but only where this will not associate the survey work with one particular sectarian division. It is usually helpful to identify local leaders and other prominent people, and to make specific efforts to inform them about the survey, and to gain their cooperation. Such people can be particularly useful in spreading information, and may even assist in persuading reluctant respondents once their own cooperation is gained.

Some channels reach the general population directly – radio, television, newspapers and magazines, posters and leaflets. Radio, in particular, can be a wide-reaching medium, particularly if the survey publicity is inserted into, or next to, a popular programme. As regards all these media, publicity should be disseminated in all locally important languages.

If the survey work encounters hostility, one must discover why immediately. This is not easy – usually people are reluctant to discuss or even admit to hostility or non-cooperation. It is worth looking into the respondent population's previous experience of survey work, or of contacts with research bodies, as this is a likely source of the difficulty. Other possible causes of hostility are political, ethnic or religious differences between the local population and either the national government or the institution doing the survey, or the survey's inappropriate coverage of sensitive topics. Once the source of the problem has been identified, it needs to be dealt with either by more publicity, directed specifically at the problem issues, or by modifying the survey contents, approach or universe.

Before the enumerator begins to interview the individual respondents, he or she must be introduced to the whole local community. This is best done at a community meeting, where the enumerator and supervisor can be introduced by the local leader, and the supervisor can explain the purpose and nature of the survey.

It is usually effective to use a mixture of these various channels of publicity, and to combine them into a programme which builds up to a peak at the start of field work, as in Example 8.5.

EXAMPLE 8.5

Example publicity schedule

March 2	**Meet Chief and advisors** **Meet Local Government Council**
March 4	**Put up posters** **Meet Farmers' Association** **Distribute leaflets**
March 6	**Meet Residents' Association** **Article to appear in *The Farmer*** **Distribute leaflets**
March 8	**Radio interview in 'Farming Today'** **Article in *Daily News*** **Meeting with school teachers** **Meeting with priest** **Meeting with large farmers**
March 10	**TV interview in 'Today'** **Radio interview in 'Newsday'**

	Loudspeaker vans out
March 12	**Loudspeaker vans out** **Start direct contact with individual farmers, and enumeration**

When interviews begin the enumerator should be introduced at the first interview with each respondent by the local leader – the traditional community head or local government or administration representative. At this initial visit he should also be accompanied by the supervisor.

The first contact between the enumerator and the respondent is of crucial importance in setting the entire tone of the relationship, with all the consequences this implies regarding the degree of cooperation and the quality of data which may be achieved.

The first impression received by the respondent is of the physical appearance of the enumerator, who should therefore be neat and clean, but not overdressed or 'too smart' for the environment in which he is working. Coupled with this visual impression is that given by the manner and attitude of the enumerator. It is important that the enumerator follows properly the appropriate courtesies of greeting and introduction within which he or she is operating (probably rural and conservative), including the correct degree of respect for the respondent's age and local status.

It is advisable for the enumerator to discuss with the respondent the likely time to be taken for the enumeration, and if necessary to arrange to return at a more convenient time. It is also best to make clear at this point any subsequent visits which will be necessary, and any measurements or other special activities which will need to be done.

For some kinds of information, the location of the interview becomes important, particularly whether it is at the farm land, in the home, or in some public place. It is also necessary to consider whether it is preferable for the respondent to be interviewed alone and privately, or with other household members or neighbours (or even casual passers-by) present. These issues were discussed in detail in Chapter 4.

Respondents often ask for money or other payments in return for cooperating in a survey. Other than in very exceptional circumstances, for example involving a large input of time or work on the part of the respondent, it is not generally feasible or desirable to make any payments. In a typical, fairly large-scale survey, the numbers of respondents are too large for the funding of any substantial payment to be possible – and this is an argument which is often acceptable to respondents. Small gifts such as pencils and notebooks for children, or pictorial calendars, may be appropriate. In many societies, photographs of the respondents are popular gifts.

Respondent selection

As was discussed in Chapter 3, for many surveys a multi-stage sample design is likely to be appropriate. For such designs, and also for small single-stage surveys without previously available frames, it is necessary to construct the last-stage sampling frame as an early part of the survey operation itself. In this type of survey, the sampling unit is likely to be a household, and normally the most convenient way to construct a frame is to number and list dwellings, and then list households within dwellings. Where dwellings are irregularly laid out and not already numbered in some way, it is difficult both to ensure that none are missed or repeated, and to identify easily a particular dwelling during subsequent enumeration work. In this situation some form of physical numbering of the dwellings, perhaps with paint or chalk or sticky labels, is necessary.

In preparation of the frame, each household should be listed by dwelling/household number, together with some other household identification, typically the name of the household head. Additional information for further selection or stratification may also be included. This subject was discussed in more detail in Chapter 3.

If the sampling unit is not the household, then a different approach to preparation of the sampling frame is required. If the sampling unit is a holding, a parcel of land, or an area of cultivation, then in order to create a frame it will be necessary to produce a sketch-map of all such units in the survey area, and then number and list them in as systematic a manner as possible. Again, it may be necessary to include in the listing some subsidiary information to use for selection or stratification, for example the crop grown, the area, or the cultivation status.

Generally speaking, the most convenient method of selection of respondents is to use the serial numbers in a listing of the sample units of the kind discussed in the previous paragraph. At this stage, the method of sampling is most likely to be either simple random sampling or linear systematic sampling (as described in Chapter 3), which require a set of random numbers for selection of all the respondents, or a random start number and an interval, respectively. These numbers should preferably be selected and determined by senior staff at headquarters, but it is often convenient for the selection itself to be done in the field. In this case, it is advisable that the details of the selection procedure, particularly the random numbers to be used, should not be known to the field staff until after the sampling frame is constructed – otherwise, it is very easy to juggle the frame listing so that the selected sample units are particularly easy ones to enumerate.

The sample frame and sample selection listings and related documents should be retained and stored not only during the complete enumeration exercise, but also permanently with the survey forms and other documents. This is advisable for purposes of checking and verification, especially if the survey produces unexpected results, and it is useful to be able to investigate the possibility of sampling biases. Also, a sampling frame is an important resource which may very well prove invaluable for later survey exercises.

In a multi-round survey, or one involving several visits to obtain information relating to different seasons or seasonal operations, it is obviously necessary to make provision for subsequent contacts with the respondents. The programme and timing of the subsequent visits, and the information to be gathered at each one, need to be specified at the planning stage of the survey, and all the organisational arrangements made accordingly.

The survey lists and records should contain adequate information for each respondent to be easily traced for the subsequent interviews, and for the correctness of the tracing to be checked, for example by repeating some standard questions such as details of the household members, and checking that the responses match with those of the previous interview.

Often some respondents will no longer be available at subsequent interviews, having perhaps moved away, died, or become ineligible for the survey (for example, by a farmer selling his cows in a survey of dairy farmers). A set of decision rules should be formulated beforehand, to be applied in such circumstances. What these should be depends very much on the individual survey, but they should be included in the survey manual, and the enumerators trained in their application.

Field supervision

Supervision to check that the enumerators are doing their work, and are doing it correctly, needs to be done both in the field and in the supervisor's office. First of all the supervisors should make random visits to the field to check that the enumerators are working according to their programme. At the same time, they should observe the enumerators at work, and check that they are following correct procedures of data collection, whether in measuring, observing or interviewing. The supervisors should also re-check a sample of each enumerator's work, by returning to a measurement or interview previously recorded, repeating it themselves, and checking whether their results correspond with those of the enumerator. Any discrepancy found may not necessarily be due to inadequate work on the part of the enumerator – it may for example reflect a failure of memory on the part of the respondent – but all discrepancies must be very thoroughly investigated, and any necessary corrective re-training, or even disciplinary action, should be dealt with immediately.

Once reasonable quantities of data have been gathered, office-based checking is as necessary as field checking. It is at this stage that the supervisor has the opportunity to assess the overall consistency and plausibility of the data gathered. Any persistent errors or misconceptions can be identified, and corrective action taken. Patterns of suspiciously similar or even identical data gathered by any particular enumerator can be observed and investigated, to assess whether data fabrication is taking place. Consistently different ranges of results from different localities or from different enumerators can be identified and investigated. In general, this stage of checking is aimed at implausible or inconsistent patterns of data, rather than at individual items. This stage of checking is essentially unstructured, and the most likely kinds of problems are ones which cannot be anticipated.

The response to errors should be as rapid as possible, so that their effect on the data gathered can be minimised. At least a weekly check on each enumerator, and subsequent rapid follow-up of errors, is therefore essential.

When an error is identified, it must of course be corrected – but even more importantly, steps must be taken to ensure that it does not occur again. If the error found is an isolated instance, then some simple on the spot re-training that emphasises the need for care in enumeration is appropriate. If the error is a persistent one, reflecting some misunderstanding of a question or concept, then more detailed re-training relating to this question is necessary. In this latter case it is advisable to check for the same error with other enumerators, in case the misunderstanding has been repeated elsewhere. Also, special attention to this question during subsequent checking is appropriate. One of the functions of the supervisor is to ensure consistency between the work of the different enumerators. Where wrong information on a question has already been collected for part of the sample, then either that information must be gathered (correctly) again, or else the question must be dropped entirely. Which of these is chosen depends on:

- how important the question is
- how large a part of the sample has already been wrongly enumerated
- how costly in resources and time it would be to go back and gather data again.
- whether it is physically possible to repeat the measurement or interview

The surveyor must assess the comparative costs and benefits of the two alternatives and choose one – there is no middle course.

In extreme cases, where an enumerator is persistently error-prone, and is either incapable of improving his work or unwilling to take the trouble to do so, then dismissal must be considered. In case this becomes necessary, it is advisable to recruit and train additional enumerators, and to select additional units for inclusion in the sample, so that the damage can be minimised.

Other kinds of errors, as well as those of enumeration, may be found during the fieldwork. It may be discovered at this stage that the sampling frame is deficient, with either serious omissions, or errors in stratification data. In the case of the first of these problems, it is necessary to construct the missing portions of the sampling frame, and to do any extra sample selection necessary. In the case of the second, the options are:

- correct the errors, and if necessary re-select the sample
- correct the errors, but retain the sample already selected, and weight the strata if necessary
- drop the stratification altogether

Again, which option is chosen depends on the surveyor's assessment of the relative costs and benefits of each course of action, in any particular instance.

At the main survey stage it may be found that the work at the preparatory and pilot survey stages was deficient or incomplete, and some of the survey questions were wrongly-framed to produce good information, or perhaps too sensitive or unacceptable. Such questions must then be dropped from the survey. Whether they are replaced by better or more appropriate questions (which must then be put to all the respondents, including those already enumerated) depends on:

- how important the topic is
- whether good information is likely to be obtainable from a modified approach
- how large a part of the sample has already been enumerated
- how costly in resources and time it would be to insert a modified question in the survey

In general, anything which is not satisfactory or working properly has to be dropped from the survey. It may or may not be replaced by something more satisfactory, depending on the surveyor's assessment of the costs and benefits, and of the constraints upon it.

Pre- and post-enumeration field work

The quality of information obtained can be much improved by a good pilot study. A major purpose of this is to check whether the organisation and arrangements of the survey actually work satisfactorily. It is not sufficient simply to test the questionnaire for comprehension and appropriateness on a few trial respondents – the whole of the survey operation in all its aspects must be tested out, albeit on a small scale. This approach thus checks the administrative and organisational arrangements in general, the arrangements for the supply and distribution of all the resources and equipment needed for the survey (as discussed above), as well as the fieldwork operations, the survey forms and manual, and the data processing arrangements.

The pilot study should thus proceed through all the stages and operations of the survey proper, but on a small scale in a few selected localities. These localities should be chosen to cover as complete a range as possible of the types of area and population to be covered by the survey. Where relevant, the pilot study should include areas of different topography, soils, climatic conditions, agricultural practices and cropping patterns, settlement patterns, and ethnic groups and cultures. But this is the ideal, and there may not be enough resources or time to cover all the different areas – in this case the priority is to cover a few areas but over a broad range of characteristics. Areas should not be excluded on grounds of difficulty of access (unless, indeed, they are to be excluded from the survey proper). If any of the survey arrangements prove to be unsatisfactory it is best that this be discovered at the pilot stage, rather than later.

Since the purpose of the pilot study is to identify weaknesses and problems with the survey materials, procedures and arrangements, the senior technical staff should be closely involved in the pilot study. The survey forms and procedures must be observed under operational conditions in the field if problems are to be correctly identified, and appropriate solutions found. Thus all the senior level technical staff should be in the field during the pilot study, and should observe all stages of the work as it is being done under field conditions.

If the pilot study is properly done, it is likely to lead to changes to the survey forms and manuals, and to the procedures and organisational arrangements. It is therefore necessary for the pilot study to be timed to allow enough time to analyse the results and observations from it, decide upon amendments where necessary, and produce revised materials and arrangements in good time for the start of the main survey operations.

At the other end of the field work, for most surveys, a post-enumeration survey is the only means of obtaining any objective and quantitative assessment of the non-sampling error. As discussed in Chapter 3, the non-sampling error is generally at least as important as the sampling error in a survey, and is likely to be increased if the scale of the survey is expanded to reduce the sampling error. However, while the sampling error can fairly readily be estimated, and attention can be directed to its effect on the precision of estimates, any assessment of non-sampling error is usually conspicuous by its absence from survey reports. Apart from the use of inter-penetrating samples (also discussed in Chapter 3), which for many types of survey is logistically and organisationally impracticable, the only way of calculating this error is by conducting a post-enumeration survey.

A post-enumeration survey (PES) is different from a survey proper in a number of respects, because its purpose is not strictly speaking to collect information but to assess the degree of correctness and accuracy with which the existing data has been gathered. The PES should therefore be conducted under the most favourable conditions possible for gathering accurate information.

This is commonly achieved, or at least attempted, by conducting the PES on a random sub sample of the complete survey sample, and using more senior, supervisory staff for the enumeration. It is also often appropriate to select a subset of the questions for the PES, choosing the most critical items in the survey for this assessment. This reduces the work of the PES itself and hopefully increases the accuracy. For example, in a livestock survey, the PES might concentrate on just the numbers of the different stock types. These reductions in scale of the PES also assist in reducing the time and resources needed for processing and analysing it.

The number of respondents to be included in the PES will depend upon the expected level of non-sampling error arising from observations during survey supervision.

The major potential pitfall in these arrangements for the PES is that the more senior staff conducting it may assume that because of their greater seniority, expertise and experience, they will automatically attain a greater degree of accuracy in data collection. They may also assume that the work is at a notionally 'lower' level than what they are accustomed to, and is demeaning. The work may therefore be done carelessly, and without the attack and rigour that is needed for this exercise to obtain high data quality. To counter this, it is advisable at the preparatory stage to discuss the need for and importance of the PES, and the necessity of having senior staff conduct it. It is also necessary to involve the most senior levels of technical and professional staff in this work, not just the field supervisory levels.

In a survey with multiple visits it is particularly difficult to organise a PES. Here a stage of the PES should follow as closely as possible to each stage of the actual enumeration, if the same data are to be gathered with any reasonable prospect of accuracy. To reduce the burden of PES work on individual enumerators and respondents, and to remove the possibility of enumerators giving particular care and attention to the PES sample once they know what it is, it may be helpful if each stage of the PES is done on a different random subsample, where the nature of the data collected allows this.

For all types of PES it is vital that the subsample is selected on a random basis, so that the resulting estimates of non-sampling error are applicable for the survey itself. It is also necessary to ensure that field supervisors are not allocated survey areas which they were responsible for supervising during the main survey, otherwise the PES cannot be regarded as independent of the main survey, and again the resulting estimates of non-sampling error may not be generally valid.

Note

1. Few texts have been devoted to the implementation and management of survey programmes. Casley and Lury (1981) devote a chapter to the survey team and another to data collection. Collinson's East African experience (1972) gives a good account

of the practicalities of data collection in a formal setting. Circumstances vary widely and to avoid a rigid, simplistic set of prescriptions the reader should seek out accounts from as wide a range of studies as possible – rapid appraisal, small-scale surveys, national surveys, and social or anthropological studies. Suggestions include Barnett (1977); Srinivas *et al.* (1979); McCracken (1988); Norman (1970).

9

Data processing

9.1 Organisation

The objective of data processing is to prepare raw data for statistical analysis and presentation. It is part of the sequence of data transfers from an event in reality to a record suitable for analysis. As such, data processing brings together procedures for ensuring data quality, by means of cross-checking and inspecting information on the survey forms, and the preliminary stages of analysis. These activities are in fact mutually supportive, and it would be a false distinction to try to set them apart.

For all except the smallest of surveys it is beneficial to undertake processing and analysis in a sequence of increasingly complex stages. The first step with the data is to produce some simple summaries of the variables under study, either for the survey as a whole or for specific villages or regions. These summaries will enable a rough but effective assessment of the basic quality of the data, of the magnitude of the results compared with prior expectations, and of the major features for analysis. After the summaries of variables have been produced there may be further editing of the data set, or even follow-up visits to selected respondents for additional enquiries.

The next step is to prepare a preliminary set of tables covering the main analytical topics. The tables should present the main findings from the survey with an indication of statistical precision if appropriate. It is often useful for results from this stage to be circulated with comments, for review by interested parties. The response may highlight alternative courses of further analysis or comparison with existing data.

Further analysis will depend on the purpose of the survey. If the data are being collected as part of a time series there may be a wait of several years before more detailed analysis can take place. If the survey has to stand alone, there will normally be a requirement to prepare a survey report which includes more detailed analysis and places the study in the context of similar knowledge about the same subject. Alternatively, the data may be added to an existing data-base and be used on an *ad hoc* basis when required.

Quite apart from the benefits which come from approaching the data in a progressive manner, there is also the issue of timeliness. Production of preliminary results can be achieved relatively soon after fieldwork is completed. Surveys are expensive and intrusive. Both the funding institution and the participants deserve a high priority being given to presentation of results. An analytical strategy which defers all reporting until a 'final' survey report may involve delays of many years, and almost certainly will result in a loss of confidence on the part of the potential users of the data.

The sequence of operations are independent of the medium chosen for data processing. Two alternatives face the surveyor: to process by hand or by computer. Whichever approach is chosen, the sequence follows the same pattern.

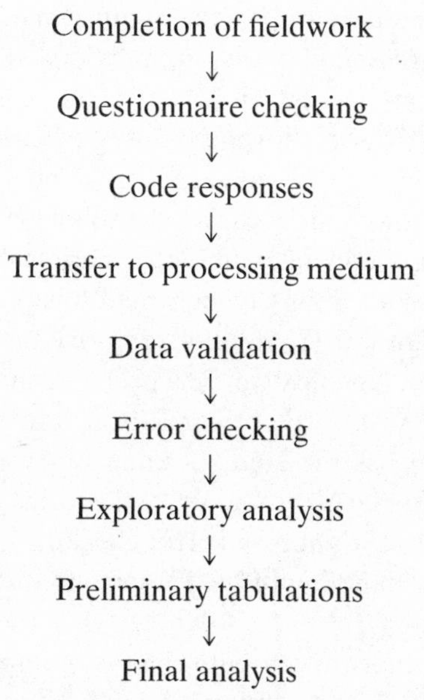

Questionnaire coding was described in Chapter 7. Sometimes this will be carried out in the field, by enumerators and their supervisors, but very often will be undertaken at the office as part of the initial checking procedures.

9.2 Preparation for processing

Survey forms

When the survey forms are returned to the office the first step is to organise them in logical sequential units, preferably bound together in a file case or box. The forms should be stored in numerical order corresponding to the 'Identification' coding chosen for the survey. Each bundle should have a covering summary form which indicates the name and code number of the sample unit, together with the number of forms or documents in the bundle. Provision should also be made for comments or notes about missing or damaged forms, and observations by office staff.

A large survey which is being handled by a number of analysts and clerks will benefit from a formal storage system, with a master location index and reference cards for each storage location. In this way every time the original forms are consulted a record is made of the person using them, the date they are taken from store and the date returned. To be effective, forms should be under the control of a nominated person who is responsible for checking their completeness every time they are withdrawn from or returned to store. In the enthusiam of exploring problems and interesting features it is very easy for individual documents to become separated from the complete set, and once lost they can never be replaced.

Few surveys have a budget for the printing of forms on heavy weight, good quality paper. Most are managed on standard office duplicating sheets, which are neither hard wearing nor robust enough to cope with a short but demanding life in an enumerator's briefcase. Spells of sun and rain, and repeated corrections to pencilled answers, quickly turn the surveyor's neat design into a tattered scrap. By the time the forms are returned to the office, legibility can be a problem and physical survival in doubt. It is important that the original forms are preserved, and kept accessible for as long as the study is under analysis. For this reason, special care must be taken in the storage of the forms, and the transfer to a processing medium. Validation and error checking must be controlled to minimise unecessary handling of the forms, and maximise effectiveness of the procedures.

Once coding and on-form computations are complete, the coded data must be transferred to a processing medium. For manual analysis two options exist. If the data volume is low, tables and statistics can be prepared directly from the forms. However, this is rarely the best approach. It is uncommon for final versions to be achieved at the first attempt and therefore every form would have to be handled repeatedly, with an increased chance of losing data and causing physical damage. The preferred approach is to transfer the data directly to a tabulation sheet. For computer analysis the data must be entered into the computer and stored in a data file. Although the two approaches differ, the principles and structure of a computer file follow closely the logic of a tabulation sheet.

Tabulation sheets

A tabulation sheet is a simple table of rows and columns. Each column is used to store the answer to successive questions on the survey form, and each form occupies one row of the table. Another description would be to call each column a variable and each row a case. More formal terms are introduced later. For a simple example consider a questionnaire about household ownership of cattle, which also records farm size, and family size. A tabulation sheet containing eight questions is set out in Example 9.1

Each column is labelled with a key word or phrase describing the question on the form, or the subject of that variable. 'Ident' stands for the identification code of the form; 'Farm' is farm size in hectares; 'Adequiv' is the family size expressed in adult equivalents. Data are entered by rows in numerical sequence. This sequence should correspond to the physical order in which the forms are stored. Unless the survey is a single-stage sample, it is convenient to group the cases within one stage, such as a village or region, on each sheet.

EXAMPLE 9.1

Tabulation sheet

Ident	Year	Farm	Oxen	Bulls	Cows	Calves	Family	Adequiv
101	86	1.43	1	1	1	1	4	3
102	86	1.21	1	0	0	0	4	3
103	86	0.95	0	0	0	0	2	2
104	86	1.35	1	1	2	2	5	3.5
105	86	1.61	2	1	2	1	6	4
	$\bar{x}$	1.31					4.2	3.1
	SD	0.25					1.48	0.74
	n	5					5	5

Tally sheet

Analysis can be carried out on the sheets themselves, or onto an intermediate document, a tally sheet. If the analysis requires counts, proportions, totals and averages of single variables, these can be speedily computed and the resulting answer written at the foot of each column. More often a cross-tabulation between two variables is needed. In this calculation, observations from two variables are compared for each case. From the tabulation sheet above, a tally sheet could be used to compare ownership of cows with ownership of bulls. The first step is to find the maximum and minimum value of each variable and draw up a tally sheet with cows along one axis and bulls along the other, as in Example 9.2.

EXAMPLE 9.2

Tally sheet

Bulls	Cows 0	1	2	3	4	5	>5	Total
0	(9)	(6)	(2)	(3)		(1)		21
1		(3)	(10)	(4)	(2)			19
2			(1)	(1)			(1)	3
>2						(1)		1
	9	9	13	8	2	2	1	44

Each case is then inspected. If a farmer owns one bull and five cows a tick (or tally) is placed along the row for one bull, under the column for five cows. Counting is made easier if the tally marks are grouped in fives using simple systems such as four strokes crossed out by the fifth. When all the data are inspected the totals are written in each cell of the tally sheet, and converted to percentages if necessary.

When the tallies are completed check that the totals for rows add up to the same grand total as that for columns. If they do not agree, a mistake has been made.

In a similar way, a tally sheet can be used to calculate averages or other descriptive statistics of one variable, grouped by another variable. In Example 9.3, farm size has been computed by number of oxen owned.

EXAMPLE 9.3

Number of Oxen

	0	1	2	>2
Farm Sizes (ha)	0·67	1·04	2·80	2·68
	1·04	1·92	6·48	2·87
	0·57	1·90	2·89	2·09
	0·28	2·07	4·42	2·29
	1·04	3·83	1·81	1·70
	1·12	1·18	2·17	1·61
	1·15	1·05	2·04	3·28
	1·87	1·21	1·75	3·06
$\bar{x}$	0·97	1·68	3·07	2·45
SD	0·48	0·69	1·61	0·62
Σx	7·74	13·40	24·56	19·58
n	8	8	8	8

The tabulation sheet (Example 9.1) we have used has just nine variables. Clearly for lengthy survey forms each record would stretch over several pages. A paper sheet may be no more robust than the original forms and many surveyors use financial ledgers or loose-leaf files for the sheets. It is helpful to complete the sheets with the units of study in sequential order, and group sampling clusters such as villages (if the design includes such a stage) on single sheets where possible. If the survey makes use of multiple questionnaires a decision must be made either to prepare separate tabulation sheets or to combine the data onto one sheet.

The advantage of a manual system is that it is simple to prepare descriptive statistics for ordered, contiguous subgroups of the data, and data can easily be transferred to tally sheets for cross-tabulation. The physical process of preparing the sheets and extracting statistics brings close contact (literally) with the data, which is invaluable for the analyst wishing to understand the content and limitations of the survey. It is a rigid medium however, and there are a number of disadvantages. The process involves a large and unavoidable number of data transfers, which are potential sources of error. Data sets commonly require the creation of computed variables such as crop yield (output divided by area) which do not appear on the survey form in the computed format. Space must be set aside on the tabulation sheet for later calculation of these variables, but it is difficult to predict all that will be needed in advance of analysis. New variables may have to be placed at the end of each record, far from related data.

Two-way cross tabulations require a simultaneous extraction of two variables. If these are located far apart on the sheet the process can be awkward and prone to errors in reading the wrong item. Much of the statistical analysis is likely not to be by sample grouping, but by stratifying the data according to observed characteristics, for example, cropping pattern by farm size, crop output by farm power, farm size by land tenure status. Organising tables by these strata involves the extraction of individual records scattered throughout the data set. Extreme care is necessary to ensure that the correct values are assigned to each stratum or class, and that observations from subpopulations are numerically consistent from one table to another.

Most critical of all for agricultural surveys, the tabulation sheet format is best suited to data which are organised in a one to one relationship. That is to say, data for which each case or record owns one occurrence of each variable. Many farm surveys take the farm household as the unit of study (despite problems with definition), but collect information about the members of that household, and about the fields farmed by members of the household. This brings complications to the data structures. Each household owns several fields. Each field will contain different crops grown under different management. but some crops will be grown on more than one field. In other words there can be multiple records for both fields and crops within the household record. Sometimes we would need to analyse by field, to examine production by location or soil type. Another time the analysis must be by crop, to determine seed rate or fertiliser use. Data from subunits such as fields have to be usable both as separate records, and aggregated to a parent record such as the household. This will affect the choice of variables for tabulation and is a significant element in the choice between manual and computer analysis. We return to this in section 9.3.

Computer data files

Computer processing involves the same principles and approach as analysis by hand. The most important differences concern the data files and data structures, which will reflect the type of software being used.

First, some terminology. Variables and cases were introduced for manual tabulation sheets earlier in this chapter. Other terms are in common use for computer data files. The terminology differs both between software packages and at different stages in processing. Table 9.1 illustrates this in comparative terms.

The set of data on each questionnaire is known as a case or a *record*. Every record consists of *fields* or data *variables*. A field contains a specific number of *characters* which define the *length* of each field. Some software will permit a mixture of alphabetic letters, numbers, punctuation and mathematical symbols to be entered and stored in data fields. But most software requires that fields are defined as containing specific types of characters. A typical classification would include *numeric* fields containing only numbers (with decimal points, + and − signs) and *character* or *string* or *alphanumeric* fields, which may contain any letters or numbers, and often other symbols as well. The software being used may require more specific definition of variables, and also permit different types of variables. Numbers sometimes have to be defined as *real* (with decimal fractions) and *integral* (whole numbers only). Many data management programmes include *logical* variables, with values representing true and false, and *date* variables.

Manual tabulation sheets were described as a rigid medium. They are rigid in the sense that once the data are entered further manipulation is difficult. By comparison, computer data files can usually be manipulated easily, but the initial creation of the computer file is more demanding than for a tabulation sheet. Each field must be correctly specified in length and content. A tabulation sheet will cope with numbers of any size for each field, and where a number is missing a comment can be added to alert the statistical analyst. On a computer, provision must be made for the correct maximum dimensions of each field, for missing data and for codes or symbols to identify records with particular characteristics, when the file

Table 9.1. *Comparative data terminology*

Stage/package	One set of data	Data item
Questionnaire	Respondent or unit of study	Question
Tabulation sheet	Case	Variable
Computer spreadsheet	Row	Column
Computer database	Record	Data field
Computer statistics package	Case	Variable

EXAMPLE 9.4

Structure of a computer file created by dBase III

Structure for database: b:ox86.dbf
Number of data records: 331
Date of last update: 05/08/90

Field	Field name	Type	Width	Dec
1	IDENT	Numeric	4	
2	FMR	Numeric	2	
3	SOIL	Character	1	
4	OXEN	Numeric	1	
5	YEAR	Numeric	2	
6	LVAL	Numeric	3	
7	LDAT	Date	8	
8	FAMILY	Numeric	2	
9	ADEQUIV	Numeric	3	1
10	FARM	Numeric	5	2
11	QUIN_OUT	Numeric	5	2
12	ANNUAL	Numeric	4	2
** Total **			41	

The computer file structure sets out the details of the file. Each field is numbered, from 1 to 12. The fields are named, and in this example are allowed names up to ten characters in width. The type of each field is indicated together with its width. IDENT is a numeric field, four characters in width, with no decimal places. SOIL is a character field, one character in width, which would permit any single alphanumeric character. LDAT is a date field, which this program stores as month, day, year, in the format mm/dd/yy, and occupies eight characters. FARM is a numeric field for farm size. It is five characters long, with two decimal places. With this program that means numbers of the format 99.99 can be stored. The decimal point is counted as one of the characters. The total includes one extra character used by the program.

[dBase III is a trade mark of Ashton Tate Ltd]

structure is created. It is helpful for the field dimensions to be represented on the questionnaire or coding form, so that problems are identified before computer entry. Example 9.4 shows a data file structure from dBase III.

Data processing by computer is described in section 9.5. Procedures are specific to the software being used but the stages to be followed are the same as for manual processing. One major advantage, however, is the ability to handle more complex data relationships by computer. This problem is discussed in the next section.

Summary

Store forms in sequential units

Tabulation sheets	– row/column data layout (flexible)	Computer files case/variable layout
	– one for each level of data	|
	– do simple data summaries	defined variable length and content
Tally sheets	– cross tabulations	
	– statistics of subpopulations	

9.3 Data relationships

In our example of a tabulation sheet we considered a simple data structure in which there was one occurence of each variable on the questionnaire. Many surveys involve more complex data relationships.

Three kinds of relationships can be defined between variables: (i) one to one; (ii) one to many; and (iii) many to many. For example, (i) a credit agency could make individual loans to farmers, one at a time. Until one loan is repaid the farmer is not able to take another. Thus each farmer would have one loan, and that loan would be related to just one farmer. The relationship would be one to one (1:1). A different credit policy (ii) could permit a farmer to take additional loans as long as he was on schedule for repayments to existing loans. In this case a farmer could have several loans, but each loan would be related to just one farmer. The loan relationship would be one to many ($1{:}n$). Sometimes (iii) credit is given on a collective basis to a cooperative, which on-lends to individual farmers. Depending on the rules, each member of the cooperative could have access to several loans and each loan would be related to a number of farmers. The loan relationship would be many to many ($m{:}n$). Example 9.5 demonstrates all these three possibilities.[1]

Three simple examples illustrate these relationships in farm surveys. If the sample unit is a farm household there will be a household head, often defined as the ultimate decision maker for the family. There is one head to each household. The relationship between household and household head is one to one. Each farming household will have plots of land. Unless arrangements of communal ownership exist, each plot will be under the control of one household, but each household will have a multiple number of plots. The relationship here is one to many. The members of the household will grow crops on those plots. Unless there are restrictions imposed by the society in which the household lives, the family will be free to decide what crops to grow on each plot, and it is likely the same crop will be grown on more than one plot. This relationship between crops and plots is many to many. Examples of labour

EXAMPLE 9.5

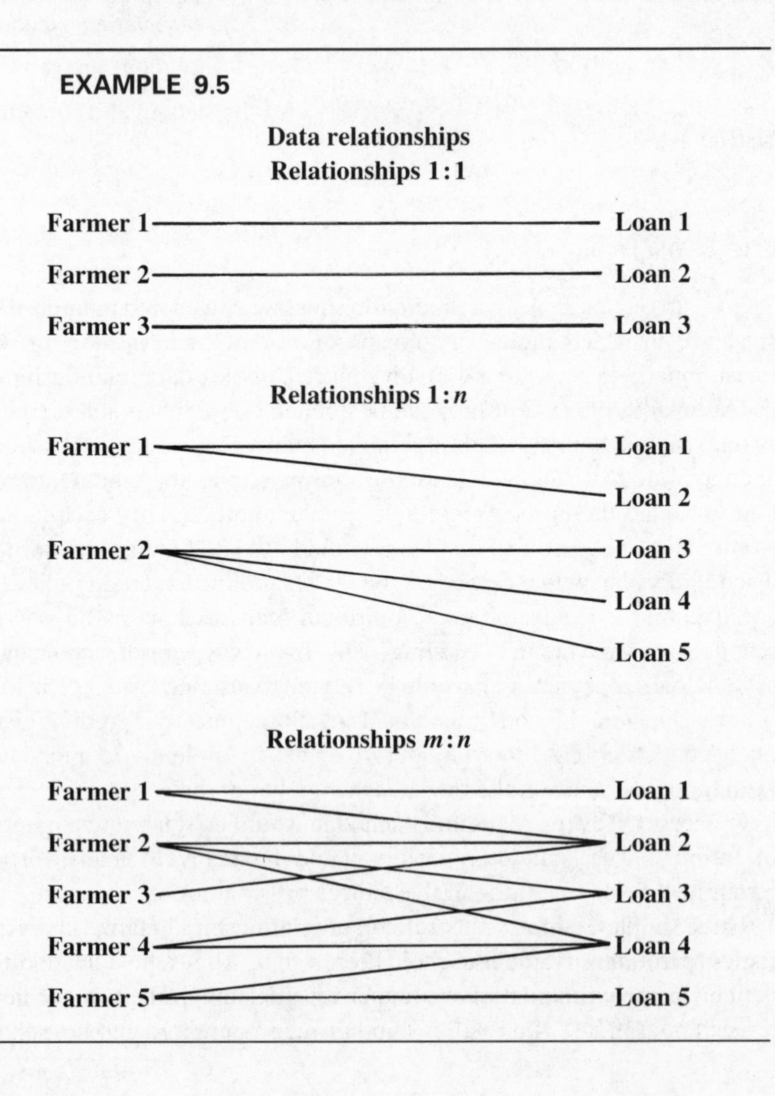

activities where members of the household work on multiple plots and multiple crops provide further illustration of complex data relationships.

These data relationships affect the approach taken in defining the layout and content of variables for either tabulation sheets or computer analysis. The solutions in both media follow the same principles, but the greater ease of manipulating data by computer permits more flexible working. Clearly, the simple tabulation layout illustrated in Example 9.1 was not designed to cope with multiple relationships. But most surveys will involve complex relations for at least some of the variables. The analyst faces important decisions at an early stage. Three choices are available.

The first, and simplest, is to make a prior decision about the variables of interest, and summarise these onto a simple tabulation sheet or computer data file. Thus for a household with a number of plots and crops, the important variables might be chosen to be the area under each crop, and the output of each crop. This would aggregate crop data across all plots within the household. Analysis by crop will then be straightforward, but any data which are plot specific, such as fertiliser input, or plant spacing or date and number of weedings, would be inaccessible, because this type of information must be related to the specific crop and area of the plot.

A second approach is to design a more complex tabulation sheet which would cater for multiple relationships, and permit analysis by the household or plot units. An actual system is shown in Example 9.6.

Each household is presented as a block composed of the individual plots, listed row by row in sequence, with a total ruled off before the next household starts. The column layout deals first with the crop details of each plot: the field number, crop mixture (coded), number of crops grown, and plot number. Then the most important crops for the area are listed in coded sequence, starting with the crop with the lowest code value (cotton, 11). For each crop is shown the planting arragement: number of stands (planting locations) and plants per 100 square metres; and the yield in kilogrammes per hectare. The totals row under each household summarises the area under that crop, and the yield per hectare. Yield is calculated as a weighted mean (see Chapter 3) for these plots where the crop is grown. With care, tabulations can be produced about both household characteristics (crop area and production), and plot characteristics (plant density and plot crop yield).

This complex tabulation system is feasible, but great care would be necessary to minimise errors. It could be undertaken on a computer

EXAMPLE 9.6

Household/plot tabulation system

Village 221 (Marke)						Cotton (11)				Cowpea (12)				Groundnut (14)				Sorghum (16)				Millet (23)			
Household	Field	Crop Mix	No. crops	Plot	AREA (ha)		Stand	Plant	Yield		Stand	Plant	Yield		Stand	Plant	Yield		Stand	Plant	Yield		Stand	Plant	Yield
	01	16, 23	2	1	0.454														241	320	1450		175		700
	02	12	1	1	0.718								546												
70	03	11, 23	2	1	0.858		184	536	532														110		400
	04	16, 23	2	1	0.674														174	280	1100		140		550
	05	16, 23	2	1	0.367														253	300	1500		160		800
		Totals			3.071	0.858			532	0.718			546					1.495			1304	2.353			563
	01	14, 16, 23	3	1	0.746										266		569		160	285	1080		140		725
71	02	16	1	1	3.135														230	360	1450				
	02	11	1	2	1.467		218	378	194																
	03	16, 23	2	1	0.511														175	290	1150		180		930
		Totals			5.859	1.467			194					0.746			569	4.392			1352	1.257			808

72	01	16, 23	2	1	2.363											120	260	940		100	1150
	02	12, 16	2	1	2.629						106					240	370	1340			
	02	12, 14, 16	3	2	1.077					30	251		261	279		135	250	1210			
	03	11	1	1	0.308			1385													
		Totals			6.377	0.308		1385	3.706		148	1.077		279	6.069			1161	2.363		1150
80	01	16, 23	2	1	1.324											180	240	1600		150	1310
		Totals			1.324										1.324			1600	1.324		1310
81	01	16, 23	2	1	0.070											105	290	800		80	650
	02	11	1	1	0.429			224													
	03	16	1	1	1.112											140	330	1370			
		Totals			1.611	0.429		224							1.182			1336	0.070		650

spreadsheet, as described below, with the advantage that calculations could be automated and hence less prone to careless errors. It would be cumbersome for large data sets but could be attractive for a case study where close physical contact with the data is desirable.

Both the simplifying aggregation and the complex tabulation share a common fault. A prior decision has to be made about the content and physical structure of the data. Unless the survey contains only a small number of variables, some items will be left out and some relationships not shown. This means that the scope for analysis will be limited. It is limited by the physical arrangement of the data, and the problem is called *physical data dependency* (Robinson, 1981: 18). It means that future uses to which the data can be put are constrained by their physical arrangement.

The third approach, which is best suited to computer processing, is to arrange the tabulation sheets so that all data relationships are portrayed on a one to one basis. Taking our farm household again, there would need to be one tabulation sheet or data file for household-level data, a further one for people-level data, one for plot-level data, and so on.

This permits maximum flexibility for future analysis, but is physically awkward for analysis between different levels by hand. One way around the problem is to ensure that every data record shares a common identification code with its parent household. In this way cross tabulations are possible. But still the system is clumsy, and many errors could be expected in transferring data between multiple tabulation sheets. By computer, however, records from different data files can be linked on the basis of a common identifying code to create specific data files as required, and complex analysis between different levels of data readily performed. Example 9.7 shows multi-level data files which can be linked by common identification codes (IDENT).

A compromise for hand analysis is to place data from higher-order levels (e.g. parent household) in front of each lower-order (plot or family) record. Thus if a household had five plots, the household-level data would be repeated five times, once next to each plot case on the plot tabulation sheet. This solution would permit tabulations which link plot and household data but at the expense of multiple repetitions of data, which would bring greater scope for error in creating the tabulation sheets, and would perhaps rarely be used. As so much of this data would be effectively redundant, this problem of repeat storage is known as *data redundancy* (ibid: 21). It is undesirable for data storage and retrieval, but may be unavoidable for computer statistical data analysis which requires a record-based data structure. In Example 9.8 the multi-level data from Example 9.7 have been combined using repeat storage. The FAMILY SIZE and FARM SIZE from the household file, have been stored next to each set of crop information from the fields file. Thus each field record can be tabulated with the corresponding family or farm size data.

Manual tabulations are an effective way of handling some data sets, especially if data volumes are small and relationships simple, but for a fully comprehensive approach to analysis computer processing is preferable.

Summary

Relationships	– physical data dependency a problem with manual storage
1 to 1	– tabulation sheets (may need several)
↓	– analyse with tally sheets
1 to many	– separate tabulations or repeat storage
↓	– link with common identifying code
many to many	– database storage in multiple files – common identifiers
	– extract data for statistical analysis

9.4 Data checking

At every stage in the process of transferring data from reality to a form suitable for analysis it is necessary to check the data which are being transferred. There are three aspects to this checking. The first is *data verification* – an examination of completeness and consistency as the data are moved from one medium to another. Next is *data validation* – to see that data items satisfy criteria of magnitude and logic. Third is the close scrutiny required to *spot fabricated data*. We shall examine these in turn.

Data verification

Verification starts in the field, and this point was touched upon earlier in Chapter 8. The first step is to ensure that every questionnaire is completed and that every question on the form is answered. This is especially important if parts of the form do not apply to every respondent. Where appropriate, each question should include provision for a response of 'Don't know' or 'Not applicable'. If these are not catered for it may be impossible later during analysis to give an appropriate coding to the response. A positive answer to each question is also necessary to show that the enumerator covered every topic. If there are no codes for 'no

EXAMPLE 9.7

Multi-level data files

(*a*) Parent level (household file)

Ident.	Family size	Farm size	No. of fields	Total output
101	5	2.47	3	15.35
102	6	2.25	4	22.15
103	2	0.95	2	15.25

(*b*) Sub-level (family file)

Ident.	Name	Relation to head	Age	Sex	Educ. level
101	John Tembo	Head	34	M	3
101	Julia	Wife	32	F	1
101	Margaret	Wife	20	F	1
101	Amos	Son	13	M	0
101	Lucy	Daughter	2	F	0
102	Benson Kaira	Head	47	M	1
102	Mary	Wife	40	F	1

(*c*) Sub-level (fields file)

Ident.	Field no.	Field area	Crop grown	No. plants	Output
101	1	0.47	Maize	241	10.4
101	2	0.90	Cotton	300	3.7
101	3	1.10	Groundnuts	325	1.25
102	1	1.00	Maize	210	9.25
102	2	0.25	Cassava	105	–
102	3	0.66	Cotton	278	1.50
102	4	0.35	Maize	200	11.40

Common identification codes

response' it will not be possible to distinguish missing data from zero values.

When the data are processed they are transferred from one document or recording medium to another. The two principal types of transfer are the initial data entry, to a tabulation sheet or a computer data file, and subsequent transfers of parts or all of the data from one sheet or file to another. Each transfer will give rise to a number of possible errors.

- some data may be lost
- some data may be repeated
- the values of some data items may be changed
- changes may be made to groups of data

During data entry, the first three types of error are possible. All are due to errors on the part of the clerk or person responsible and all can be spotted, but only through the time- consuming process of checking the entered data against the original forms.

A number of methods for checking are available. Depending on the type of data entry being employed it may be possible to enter every data item twice, independently by different people, and compare the two versions for inconsistencies. This method was in widespread use for computer data entry via punched paper cards. The data would be entered

EXAMPLE 9.8

Repeat storage of data

Ident.	Family size	Farm size	Field no.	Field area	Crop grown	No. plants	Output
101	5	2.47	1	0.47	Mz	241	10.4
101	5	2.47	2	0.90	Ct	300	3.7
101	5	2.47	3	1.10	Gnt	325	1.25
102	6	2.25	1	1.00	Mz	210	9.25
102	6	2.25	2	0.25	Cv	105	–
102	6	2.25	3	0.66	Ct	278	1.50
102	6	2.25	4	0.33	Mz	200	11.40
103	2	0.95	1	0.45	Mz	225	15.25
103	2	0.95	2	0.50	Cv	90	–

once and holes punched in the card. The set of cards would then be entered a second time, by a different operator, with the card punching machine set to an electrical verification mode. Any errors would be identified by the absence of a punched hole in the correct location, and the individual data record re-entered.

The same principle can be used with direct computer entry to a data file (key to disk) or in the completion of a manual tabulation sheet. Alternatively, a data listing can be compared visually with the original forms. This is best done by people working in pairs. One reads the correct data from the survey form while the other checks entries in the listing.

Data checking is a repetitive activity. Unless it is carefully managed, bored clerks will cause more errors than were present in the original transfer. To minimise consequential errors the work should be undertaken for relatively short periods at a time, and if possible, organised on a team basis with each team encouraged to out-perform the others. It is strongly recommended that the surveyor in charge of the survey, and his or her professional colleagues, each check at least one subgroup of the data set personally. There is no substitute for direct contact with the data. It brings a level of understanding about the strengths and weaknesses of the data which cannot be obtained in any other way and will often give rise to new ideas and suggestions for analysis and future data collection.

With very large data sets (in excess of 500 records), it may not be practicable to check every record. A formal sampling scheme can be used to check a known percentage. Every time an error is found complete checking is used for all forms originating with the same enumerator and clerk, until error-free records are found. But even with very large data sets, some variables may be considered sufficiently important to merit a complete check of every record. The time taken will be repaid by the confidence of knowing every value has been transferred correctly.

During subsequent data transfers all four types of error mentioned above are likely to occur. One of the most common reasons for transfers is to restructure part or all of the data for a particular analysis. This could involve for example, the creation of a subset of villages, or the computation of output values by merging a physical production file with price data. Either approach involves physical separation from one tabulation sheet to another, or logical separation from one data file to another. Lack of care in specifying the data items to be transferred can lead to incomplete records, repeated values, and incorrect transpositions. In the worst case, where data fields are mis-specified, whole classes of data items can be incorrectly changed.

Most modern software packages permit variables to be identified by name. But some programmes require first that the location and dimensions of the data field are identified with reference to its character position along the data record. It is very easy for this statement to be inaccurate. An error of one character position will result in a tenfold change to a ratio value, and a complete misrepresentation of an ordinal value. (See the definitions of ratio, ordinal, nominal and interval data in the glossary, and in Chapter 10.)

Two types of checks should be used to guard against these errors. First is a visual inspection of a listing of one subgroup of the transferred data, to be compared with the original listing, or a manual computation of the transformation. Second, from the whole data set, totals and counts of variables should be used to check consistency. Totals and counts are particularly important. Once the main data set has been verified and validated to a 'clean' condition, a record should be kept of the number of observations, and totals of data fields, for each subgroup such as village or region. Whenever subfiles are created, or new variables formed, the totals within subgroups for the original data fields should remain constant. This check is particularly valuable when selections are being made for statistical analysis. It is easy to mis-specify the conditions for defining subpopulations and a check of subsample size and totals will quickly reveal the error.

Data validation

Direct checking

Data validation is concerned with the correctness of the data collected. There are various approaches to exploring correctness. The most obvious, that of direct checking by repeating the interview or measurement, has limited application. One system is to check a subsample immediately after the survey. This post-enumeration survey, described in Chapter 8, is designed to check the quality of enumeration by comparing the original form with the later version. For short, single-visit surveys it is a valuable method. But it has a limited use. Many rural surveys take place over an extended period so that data about seasonal activities can be recorded as they take place. By the time the season is over, data items would have to be collected by memory recall, rather than direct observation and interview, so comparison of the post-enumeration check with the original survey is not entirely valid.

Re-checking interviews during the course of the survey, within a day or two, is possible and forms an important aspect of the work of field supervisors, but a more extensive re-checking procedure would place heavy burdens on manpower and is probably not the best use of resources. Thus validation is effectively restricted to some form of indirect checking.

Indirect checking

An item of data can be checked against three things: that it lies within a permissible range of values; that its order of magnitude or response category is plausible; and that the response is consistent with earlier observations on the same or previous interviews or with comparable independent data.

A permissible range of values can be defined both for coded and actual data. Coded response categories to multiple choice answers, and coded values for responses, both implicitly bring limitations to the range of responses. Thus in the questions in Example 7.6 the range of responses is limited to the values 0 to 6. It is a simple matter to examine this data field by hand or computer and select all records which fall outside this range. Many computer programmes include the facility to set up valid range checks during data entry. The crop code list illustrated in Example 7.8 limits the codes in a more complicated way. A valid crop code would have to lie in one of a series of ranges defined as 100–105, 110–115, 120–123, 126–129, 130–141, 150–199, etc.

Interval and ratio data can sometimes be treated in the same way. The number of days in a week or month and the reference year of the survey are numbers which lie within absolute range limits. But judgements about realistic or plausible values such as the number of children in a family, or the yield of a crop, cannot be defined in such a way, and are dealt with in the next section. It is important to note that tests of valid range do not confirm that the answer is correct. Unless the respondent's original answer is given next to the code the coded value cannot be challenged. A valid range test merely confirms that the answer is permissible. The code itself could be incorrectly assigned.

Plausibility is a harder concept to apply, and prone to pitfalls. Plausibility judgements can be made about many variables: age of children compared with age of mother; maximum crop yield per unit area; market prices for commodities, and so on. However, there is a great danger that such limits reflect an arbitrary value judgement, rather than observed variation in the population. Plausibility limits should be used only to stimulate cross checking of the data, and never to reject data without further, corroborating evidence.

Internal consistency is straightforward to identify and define. If a crop is being weeded it must have been planted; if a plot is fallow no crops can be growing on it; if a family member is engaged in a non-farm occupation some income should accrue from it. With care a questionnaire can often be designed so that the same information can be identified by two different routes and the consistency between the sources checked. Similarly the layout of sequential activities can be used to give a simple visual check of missing data items as illustrated in Example 7.13.

External consistency is more difficult to apply. It refers to the relationship between the current set of data and data collected at a different point of time, either from the same sample or the same population, or from a population which is believed to share similar characteristics. If the comparison is between the same sample at different points of time, for family size for example, physical inspection can be made against earlier records. This is something for which computer processing is ideal, but which would be very tedious and prone to error by hand.

Spotting fabricated data

Regrettably, no matter how carefully enumerators are selected and trained, from time to time fictitious data are submitted from the field. The extent of this problem can only be guessed at, but in the experience of the authors it is sufficiently widespread that the surveyor should assume that part of every survey is fictitious unless proven otherwise. There is a limit to the number of checks that can be carried out in the field. Certainly, when using teams of enumerators, a sample of interviews from every enumerator must be checked with the original respondent, by a supervisor. But many data items cannot be re-measured. Crop production is a particularly important example. Once harvested and removed to store, the chance of re-measurement is lost.

The best protection against fabricated (and inaccurate) data is regular field supervision by senior members of the survey team. Data checking is the responsibility of every officer from the most senior professional to clerks and enumerators. A surveyor who does not go to the field to check the collection of data will never have good quality results.

When the data are being processed there is relatively little that can be done to identify the problem. However, there are a number of checks that can be made based on the degree of variation in the data. A consistent feature of fictitious data is that they rarely present as much variation as is found in the real world. Examples are most common with heavy, repetitive work, such as crop harvesting. A typical situation is where an enumerator has to weigh harvest units to calculate a mean unit weight. One or two are weighed and the remainder, up to the required sample are fabricated based on the results from the two. A second example is in plot measurement. Tape and compass surveys commonly result in an incomplete polygon when the sides and bearings are plotted. If a large closing error means the enumerator has to return for a fresh survey, he can avoid the extra work by calculating the required correct final side from a graph plot, rather than using his actual measurement.

In the first example, experience shows that made-up data tend to be clustered around the mean. In the second example, the guilty enumerator would probably have a greater proportion of zero or very small closing errors than his companions. The features that can be used to search for these types of problems are the level of variation in the data, and the probability of specific occurrences.

The best simple check of data is the visual, eyeball review, when entering data or working with listings. Some people develop a knack of spotting repeated sequences, or subgroups that seem different from the rest of the survey. Unfortunately, such a skill is hard to learn if not naturally present. For most of us the first step is during exploratory data analysis. A simplified, but reliable, rule-of-thumb is that many data items from smallholder populations display standard deviations with a magnitude around 40% of the mean. Key variables, such as crop output, should be summarised by enumerator, and distribution statistics calculated. Very low, or very high, coefficients of variation do not prove there is a problem, but are a prompt to explore the data further.

The second step is to prepare frequency distributions of the suspect data. Table 9.2 presents three actual frequency distributions from surveys in West Africa. They illustrate specific types of data fabrication.

The data are crop cuts from 100 square metre subplots, weighed to an accuracy of 0.1 kg, equivalent to 10 kg/ha. Look first at List 1. At an accuracy of 0.1 kg there are 110 possible yield values between 3.4 kg and 14.3 kg. Intuitively it seems unlikely that one of those values 11.2 kg should be repeated five times, thus occurring in 15% of the cases. List 2 shows a more sophisticated fault, with the repetition of values possessing some common factor. Twenty two out of the 28 values are exact to the nearest kilogram. Only 17 of the 27 non-zero values are different from each other. Again, these are intuitively unlikely proportions. The third list does not contain repeated values, but there is a high degree of regularity in the decimal value. Not one of the yields has a final digit falling between 6 and 9, and 3 never appears. Given that one might expect an equal probability of any number 0 to 9 falling in the decimal place, the frequency distribution is improbable. But the question with each example is how improbable?

From the three examples it is the first one, repetition of a single value, which is the hardest to demonstrate as statistically very improbable. This is because it is expected that a yield distribution will have some degree of

Table 9.2 *Data fabrication – unthreshed crop yield (kg/100 m²)*

List 1		List 2		List 3	
Value	Freq.	Value	Freq.	Value	Freq.
3.4	1	0.0	1	0.0	1
7.3	2	8.0	3	2.2	1
7.4	3	9.0	1	3.2	1
8.2	2	9.4	1	3.4	1
8.3	1	10.0	2	4.5	2
9.1	1	10.6	1	5.2	1
9.2	1	12.0	4	5.5	1
9.3	1	13.0	2	6.5	1
9.4	1	14.0	3	7.1	1
10.1	1	15.0	1	11.2	1
11.0	1	17.0	1	11.5	1
11.2	5	18.0	1	12.0	1
11.3	4	18.1	1	12.1	1
12.1	1	19.0	1	13.0	1
12.3	1	20.0	2	16.2	1
12.4	1	21.0	1	17.0	1
13.0	1	30.5	1	17.5	1
13.1	1	38.0	1	18.2	1
13.2	1			19.1	1
13.3	2			19.2	1
14.3	2			20.1	1
				20.2	1
				20.4	1
				21.0	1
				22.0	1
				25.1	1
				29.1	1
				30.2	1
				31.1	1
Totals	34		28		30

central tendency and that repeated values will be more likely to occur naturally in the middle of the distribution than at the tails. At the same time it is impossible to measure the true central tendency of a distribution containing fictitious observations. A possible approach is to compare the suspect distribution with a distribution similar to that believed to prevail in unfabricated data. The typical form of yield distribution excluding cases of complete crop failure is slightly positively skewed, but reasonably close to a normal distribution. By choosing typical parameters for different types of crop a model distribution can be simulated mathematically. The probabilities of repetitions of single values, or number of different values, can be calculated from the model.

A different technique is called for when the suspected fabrication involves selection for or against specific last digits. The last digit has little influence on where the yield observation lies in the frequency distribution. Therefore, although the probability of any particular value being repeated depends on where in the distribution it lies, the probability of it having any particular last digit is effectively uniform, being 0.1 times the number of non-zero cases. In fact, the probability follows the binomial distribution and confidence intervals can be calculated for repetition of any given penultimate digit.

A set of tables for examining these defects are included in Appendix 2 (Tables A2.6 and A2.7). All the tests are probabilistic in nature. They can indicate when a particular data set is highly unlikely to have been created by accurate recording of natural processes, but they cannot prove that it is absolutely impossible for the data to have arisen naturally. There is always the possibility of rejecting data which are in fact genuine and the first stage in any data assessment exercise using these tests is therefore to decide on the acceptable level of risk of such rejection occurring. The tables are given at 95% and 99% levels of probability.

So far as assessment strategy is concerned, the probability chosen is also the measure of the chance of incorrectly excluding an unfabricated data set. For example, if it is decided to reject data sets with a less than 5% probability of repeated values, the probability of rejecting a genuine data set is also 5%, or one in 20. The risk of rejecting genuine data must be balanced against that of accepting fabricated data and thereby invalidating the analysis. The effects of fictitious data are potentially so severe that a fairly low probability of 95% is advocated, despite the occasional injustice to field enumerators and supervisors which must inevitably result. In practice the decision to reject data would be made after studying all the circumstances, not just one statistical test.

The tests for the examples shown here are described in Appendix 2. Each of the data sets fails on at least two of the three tests for repeated values, numbers of different values and repetition of the last digit.

Summary

Verification – completeness, consistency
– repeat data entry or data transfer to check
– do simple totals, averages of data items before and after transfers

Validation – logic and magnitude
– direct checks – resurvey, often not possible
– indirect checks – valid range
– magnitude
– internal consistency

If fabrication is suspected test for:
– low variation
– clusters around mean
– repeated values
– repeated last digits

9.5 Processing of data by computer

As we have stated before, computer processing follows the same stages as we have set out for manual processing. The aim is first to store the data in a machine-readable format, and then to use it for cleaning, validation and statistical analysis. Given the widespread availability of microcomputers, many readers would be in a position to use a computer, even for their first survey. The major benefits are the ability to deal with large quantities of data quickly and accurately, and the avoidance of physical records which deteriorate in use.

Computer processing can be approached in a number of different ways according to the experience of the user and the nature of the survey. In this section we first consider the issues relating to the choice of hardware. We then describe the main types of software that are available (but leaving a review of specific packages to the computer press[2]), and then put forward three alternative processing scenarios for novices and experienced users. Lastly, we finish with a list of features which are needed in the processing software.

Hardware

Considering first the hardware, there may in some cases already be a suitable machine or machines within the survey organisation or in a related body. In this situation, it is essential to check that the hardware is suitable for the task envisaged, that it will definitely be available at the time required and that it will be, practically speaking, under the control of the survey staff for the duration of the survey work. If there is any serious doubt about any of these points, then it is generally preferable to procure the hardware required specifically for the survey programme or organisation, since problems of access to or use of the computer facilities can cause extremely long delays in processing.

The first step in computer operations is the choice of computer. The type, size and capacity of computer will influence the style of operation, the software to be used and the volume of data which can be handled. The main choice will be between a mainframe or minicomputer installation, and a microcomputer. The larger the installation the more likely that specialised staff will be required to run it. Microcomputers, by their relatively low cost and ease of operation have made computer storage and processing of data more accessible to people without formal training in computers. Commercially available software will permit virtually all processing operations to be performed on a micro, so the main decision facing a surveyor will be the capacity of the machine to handle the data set. Two aspects are important here. One is the physical storage limitations of the hardware, and the other is the data handling limitations of the software. Actual processing will be affected by the speed of computations. This is a function of the computer processor, determined by the design specification and running speed.

Microcomputers are classified by the number of binary digits (bits) which make up a logical unit, or byte. Most early micros were 8-bit machines. The byte size limits the size of memory which can be addressed by the processor. This limits the amount of data which can be read into the computer memory for processing. At the time of writing most popular machines are based on 16-bit processors. Mainframe and mini computers handle logical units of 32 bits, 64 bits or more, and the trend is for microcomputers to follow this path as the hardware technology progresses.

Data storage is on magnetic disks or tape. Disks can be transportable, in a variety of sizes and capacity, or fixed, often referred to as 'hard' disks. Depending on the physical size and density of storage a transportable (floppy) disk can contain up to 1.4 megabytes (1000 kilobytes). Hard disks for microcomputers are generally in the range of 20 to 100 megabytes but higher capacities are increasingly available. Tape storage imposes physical constraints on the speed of data access, due to the sequential nature of the medium, and is primarily used for security or archival storage, or for transporting data between locations.

In order to choose the type of computer for processing, the surveyor will need to know the size of his data set and the limitations of the computer and software. Data sets require one byte for every character stored, plus an 'overhead' amount which varies according to the software being used. The data dimensions can be calculated from the length of each record, and the number of records. Thus one hundred records, each with fields that total 25 characters in length would take up at least 2500 bytes. Only the very largest of surveys would be outside the capacity of a 16-bit microcomputer to handle. A more practical limitation is disk storage capacity and speed of processing. If a data set is too large to fit on one floppy disk, and hard disk storage is not available, processing may have to be divided between subpopulations of the survey, which may be undesirable. In this case a computer with greater storage capacity is called for.

The selection of hardware needs to be done in conjunction with software, and in consideration of the skills and experience of the staff who will operate the system. It is possible to commission purpose-written software to individual specifications, or to employ programming staff to produce this, but these are generally expensive and time-consuming options, and the systems produced are seldom completely bug-free or foolproof, so that continuing software maintenance tends to be necessary. If there does exist commercial software which is reasonably close to fitting the survey specifications, it is usually preferable to choose that option. In that case, the hardware selected should be able to run the chosen software.

A further important criterion in selecting the computer system is the ease of maintenance and problem-solving. It is wise to assume for planning purposes that some breakdowns will occur, and the ease and speed with which they can be dealt with is a very important factor. Again, it may be possible to employ one's own engineering staff, and maintain one's own workshop facilities and spares inventory, but except for very large establishments this is likely to be disproportionately difficult and expensive. A preferable option, where possible, is likely to be a maintenance agreement with a dealer or computer company. It is necessary to ensure that the company is reliable and has a good local reputation, and has technical staff, spares and facilities available locally, so that a reasonable maximum response time can be agreed upon. It is also very advisable that the company should have expertise in the specific model of computer. The availability of such a company is a further relevant criterion in the selection of suitable hardware.

Software

Survey data processing involves a number of operations which can be tackled by different computer programmes (also referred to as software packages). There is therefore a choice about what software to use. Four widely available types of software are potential candidates.

Survey analysis and tabulation packages. These are software designed to enter and clean survey data, and prepare tables of results, including descriptive statistics. Examples are USP and SNAP.

Data management packages. This class of software is designed for data management for a wide variety of uses, including information retrieval systems, accounts systems, and management systems. The programmes possess good facilities for data entry and edit, but their main strength is their ability to retrieve and manipulate data using complex selection criteria. For example, they can list all data from a specific household, or list all households with a single parent, and less than one hectare of farmland. They also have the capability to use structured programmes or files with stored commands so that processing operations can be automated. Their main weakness for the analysis aspect of processing is the limited range of statistics available. For any analysis beyond simple descriptive statistics, a statistics programme is called for. Some well-known data management packages are: dBase, Foxbase, Paradox and Knowledgeman.

Statistics packages. The main stalwarts of statistical software are SPSS and SAS. Both these programmes were originally designed for use on larger computers and have migrated downwards. They are extremely powerful, offering a full range of analysis options and increasingly good data entry features as well. Their power, though attractive, means that a substantial investment in time is necessary to master them, and for many surveys much simpler and cheaper alternatives are more suitable. Other examples of statistical package are Microstat and Statgraphics, both of which have friendly menu systems to work from.

Spreadsheet packages. Our last category is the spreadsheet. These are the best known all-purpose software for microcomputers, and are credited with stimulating the growth in financial and business-related computer use. This is because the simple layout and mode of operation makes the learning process much easier than for many other types of programme. The screen display shows a set of rows and columns, of which the computer screen is like a window forming part of a much larger working area. Columns are denoted by letters and rows by numbers. Each intersection of a row and column forms a cell, and numbers, words or formulae can be written into the cell. The user moves around the spreadsheet using the 'arrow' keys of the computer keyboard, and can enter data, or prepare calculations at any location. More advanced commands permit the user to move data around the sheet, adjust the format so that presentation is neat and concise, and enter mathematical formulae or statistical functions such as mean, standard deviation and even regression. The best known packages are Supercalc, Quattro, Excel and Lotus.

Processing operations

The surveyor can choose from a number of different approaches to computerisation. Two broad strategies can be identified:

- to enter, clean and analyse data using a single software package
- to make use of different software for different operations, so as to benefit from the features of each package

To illustrate this, we put forward three scenarios: (a) the novice computer user, with a fairly small data set; (b) the researcher, working alone and needing complex analysis; (c) the survey officer managing a small statistics office, perhaps in a ministry, or an evaluation unit.

(a) The novice computer user, with a small data set.

By small, we mean not more than about 100 cases, with up to 50 variables per case. As a novice, it is important that the user can quickly grasp the software, and feel fully in control of the data at all times. The suggested approach is to use a spreadsheet, following the principles of manual tabulation which were set out above.

The data set should be divided into groups of common data relationships. Thus, using the ideas of Example 9.7, one spreadsheet would be used for household data, one for family data and one for plot data. The data are entered cell by cell, each row being one case and each column one variable (the answer to a question). Columns would have the name of each variable written in the top row. Each case, on every spreadsheet, would include the identification code of the questionnaire from which it was taken.

Cleaning would have to be done mostly by visual inspection, but this is quite feasible with a small data set. Once the data are satisfactory, a copy of the files should be saved and stored in a secure place.

Analysis can then begin, using copies of the data file on spreadsheet. The user can work with these copies individually, or sections of data from different sheets can be copied onto other working sheets for specific analysis. A more complex tabulation such as Example 9.6 could easily be constructed. Current versions of spreadsheets also permit sheets to be linked, so that cells in one sheet can be referenced by formulae in another sheet.

Simple descriptive statistics (see Chapter 10) can be undertaken for each variable, and statistics can be produced for selected cases. The main limitation to the spreadsheet is that tables comparing two variables cannot be produced automatically, but have to be built up from basic formulae. For a first-time user, however, the visual contact with the data is reassuring, and helps an understanding of the process of analysis. Once skills and confidence grow, the data can at any time be transferred out to a suitable statistics package if the need arises.

(b) The researcher, working alone and needing complex analysis

In this situation, it is probably best to work directly with a survey analysis or statistics package. First, decide on the structure of data files from the relationships in the data. All data on a common relationship can be stored in a common file. However, if the survey is very large (>500 cases and >100 variables), the file handling may slow the speed of your computer, so it is preferable to have several files, each with up to (say) 20 variables. From many surveys, it is convenient to group variables such as crop production, animal production, use of inputs, crop disposal, etc.

The data entry module of the software may permit some checking of the data. Further cleaning can be carried out using the ideas in Chapter 10. Once a set of clean data files has been created, make a copy of the files and store it in a secure place. Analysis can then proceed, either on the basic data files, or on working files created from them. For example, if crop output and livestock output are on two different files, it may be necessary to value each output, and then create a new file with the two values, so they can be summed and then analysed.

The advantage of this approach is the simplicity of working with one piece of software, especially where the software is complex and requires practice to master it.

(c) Officer managing a survey unit

The main distinction between this situation and that of a researcher concerns the volume of work and the need to have systems which can be operated by a number of people, some of whom may not be skilled in data processing. The approach recommended here is to use a data management package *and* a statistics package.

The data management package is used for data entry, initial cleaning, and manipulation of files to combine or restructure variables. Once those operations are completed the data are transferred to the statistics package for analysis.

The advantages of this approach hinge on the power of the data management software. First of all, the packages listed above can be controlled by programmes or command files. In other words, they can be set up to run automatically, asking the keyboard operator to respond to questions such as making a selection from operations listed in the style of a menu, and directing the operator to perform various tasks. This permits the system to be used by staff who have only basic keyboard skills. Second, the data entry facilities include the option to prepare screen layouts which resemble the questionnaire or other documents from which the data are being entered. This helps to make the system more friendly for the operator to use. Third, the data entry routines can be programmed to perform quality checks on the data, so that incorrect values can be spotted at the point of entry.

Once the data are entered, simple checks for inconsistent values can be run, and the data files configured for analysis.
Transfer to a statistics programme can be undertaken directly for the most common packages such as dBase, or via an intermediate file format such as comma separated values (CSV), or data interchange (DIF). These procedures are covered in software manuals. Once in the statistics package, analysis proceeds in the same way as the second example above.

This two stage approach has been used with success by the authors in numerous situations in the Gambia, Nigeria, Somalia, Sudan, Ethiopia, Pakistan, and India. Increasing levels of computer literacy have started to reduce the need for so much automated programming, and some of the statistics packages now offer more complex data entry facilities. But at the time of writing, this is still the most adaptable approach.

Whichever scenario suits the reader's circumstances, the procedures involve similar stages:

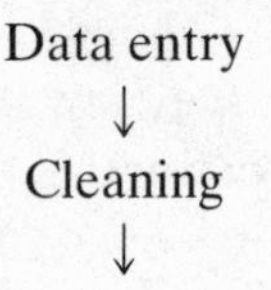

Data entry
↓
Cleaning
↓
Secure storage of 'master' copies of the data
↓
Manipulation of data to working files
↓
Analysis

For those readers who need to consider the purchase of software, we have set out a number of features which are desirable for managing the processing activities.

Data entry

- entry software should display field names and indicate dimensions
- incorrect character types should not be accepted
- valid range checks should be possible at the time of data entry
- data editing should be possible by overwriting fields listed on the screen
- it should be able to list specified records and fields to screen or printer
- there should be facilities to read and write data from and to other programmes

General

- A *recode* facility to make temporary or permanent changes to the value of observations based on their existing value, or the value of another variable, for example to create a variable of farm size classes based on the value of farm size; to combine several coded responses into a new code.
- A *missing data* provision to exclude invalid observations from analysis.
- A *transformation* facility to change existing variables, either by applying a constant or by a computation with other variables, for example to compute yield by dividing output by area; to calculate value of output by multiplying physical output by a price value. Typical transformation functions include reciprocals, logarithms, exponentiation, *Z* values, algebraic operations on vari-

ables, truncation, rounding, fractions, normalisation, scaling, and creation of a sequential variable.

- Creation of *temporary* and *permanent* new variables for specific calculations. A temporary variable is created for the current calculation only. A permanent variable is added to the data set and written to file. It is then available for future analysis.
- Ability to *join* fields from records in different files, based on selection criteria on the values of fields.
- Ability to *append* data from other files.
- Ability to *sort* the file on more than one key variable, or an *index* facility which permits a logical sort without physical rearrangement of the data records.
- An *aggregation* facility which permits a summation of values within specified selection criteria and those values to be written to a new file, for example to sum plot area from plot-based records within a household identification code, and create a new file of household plot areas.
- To choose subsets of data for analysis based on *selection* criteria on more than one variable, for example to select households in one district; or plots growing a specified crop.
- To access fields from *multiple files* based on selection criteria on the values of fields.
- To perform selections and analysis from *batch* processing or *command* files.

Summary

Computer processing

Novice – spreadsheet – transfer to statistical package when experienced
+
small data set

Independent researcher – powerful statistics software

Statistics office – database manager to enter, clean and organise
or |
Evaluation unit statistics software for analysis

Notes

1. An understanding of data relationships is necessary to design suitable tabulation sheets or computer files for statistical analysis. Analysis on its own rarely requires complex manipulation of the data, beyond aggregation to higher-order units. If the data

are to be used in their disaggregated state, however, as some form of rural database, then efficient management of relationships becomes more important. Some of the most accessible literature is the manuals accompanying software packages. For those readers wanting a deeper treatment, Robinson (1981) is recommended.

2. For example, UK magazines include *Personal Computer World*.

10

Exploratory analysis and estimation

Once the data have been prepared for computations and checked for errors, analysis can proceed. But what exactly is meant by analysis? Making sense of survey data is a creative process. General guidelines are possible but all surveys have distinct objectives and each individual must develop his or her own approach. To start with it is helpful to separate analysis into four parts.

- exploration
- estimation of characteristics
- tests of hypotheses
- quantifying relationships

Most survey analysis will involve all four aspects in differing degrees.

The first step in an analysis is to explore the data for distribution of response, central tendency and dispersion. Frequency distributions and descriptive statistics are used to identify the spread of observations and help to spot outlying values and distinctive patterns of response. For data which have been collected as part of a planning exercise, or to quantify production from a locality, the estimation of characteristics will be a major objective. Population estimates will be calculated from sample data, and reported together with an indication of the precision of the estimate obtained from the sample variance. Typical estimates are totals, such as crop areas or numbers of cattle; means, such as average family size or cultivated area per household; ratios, such as crop output per unit area or application of fertiliser per unit area; and proportions, such as the number of respondents taking credit etc. The formulae for these calculations and worked examples for three common sample designs are given in Appendix 1. The decision to put these calculations in an Appendix does not imply that they are somehow less important. But for many readers the arithmetic is inhibiting, and distracts from the main principles of analysis, which are more accessible.

Whenever a population estimate is quoted, some indication of precision should also be given. Most commonly this will be the confidence interval, described in Chapter 11. By presenting the confidence interval the reader is reminded that the sample results are only an estimate of the population and that estimates carry a margin of error.

Occasionally the calculation of population estimates will suffice. But more often the data will be used to examine relationships between variables and groups of variables. Surveys for evaluation, management and research are conducted to help guide decisions, and usually such decisions are based on an assessment of whether a characteristic differs between two subgroups of the population. The statistical approach to this is to set up a hypothesis concerning the difference between those groups and choose a test statistic which is appropriate for the estimator being used and its distribution. A typical example would be to see if a crop yield response to a new seed variety which has been observed in research trials brings the same benefit when used by farmers under farm conditions.

In this situation a test would be used to see if there is evidence of a difference between farmers using the seed and those who are not. If there is, the next step could be to quantify that difference. Benefit calculations for investment planning need estimates of the magnitude of production changes, and survey data are a source of these estimates.

Statistical techniques for tests of differences and to quantify response in the context of farm surveys, are introduced in Chapter 11 but a detailed coverage is beyond the scope of this book.[1]

10.1 Exploring the data

Exploratory analysis is the link between error checking for data quality, and analytical intepretation. Analysis of the data requires some understanding about the content of the data set. Equally, as we have seen for data checking, an assessment of the honesty of the results depends on inspection of the data items. Exploratory analysis is the last stage in looking for problems and errors, and the first stage in setting down the results. The basic objective in analysis is to describe the answers given to questions and the relationships between answers to different questions. For each question we want to know the range or *distribution* of replies given, the existence of any concentration or *central tendency* in those replies, and the *shape* of the distribution or the extent to which replies are clustered around a central point. In other words the task is to classify, summarise and explain. How this can be done will depend on the type of measurement for each question.

Scales of measurement

Four scales of measurement are commonly found in rural surveys.

- Nominal, simple classification without any expression of relative magnitude. For example, a code sequence to indicate district of birth.
- Ordinal, a ranked series, but where the distance between adjacent items in the ranking is not determined, for example, a code sequence for increasingly higher levels of education. If code 1 is primary, 2 is secondary and 3 tertiary, 2 represents a higher level than 1, but not twice as high in magnitude.
- Interval, a series which is ordered, and for which the distance between adjacent items is determined, but the zero point and unit of measurement are arbitrary, for example, temperature, time in years A.D.
- Ratio, the same as for interval but with the property that there is a true zero point, for example, the weight of production of a crop.

Most surveys make use of data which are measured on nominal and ratio scales.

Classifying the data: single variables

Frequency distribution

For a single variable the best way to start examining the data is to prepare a frequency distribution. If the data are nominal the classes are predetermined by the code sequence (Example 10.1).

A frequency distribution is also appropriate for ratio data. But the data will be continuous, rather than discrete, so an important first decision is the intervals for distribution classes. Three important rules should be followed here. First, that frequency distributions are easier to interpret if the intervals are all equal in size. Second, the intervals need to be chosen so that the cases are neither too thinly spread, nor bunched in just one or two classes. Both extremes produce meaningless distributions which are

EXAMPLE 10.1

Frequency of Response by District of Birth

Code	District	Number
1	Agew Mider	141
2	Bahir Dar	32
3	Kola Dega Damot	56
4	Mota	43
5	Bichena	200
6	Debre Markos	78
7	Metekel	10
	Total	550

From the table it is immediately apparent that there are no observations from Metekel, and relatively high numbers are reported from Agew Mider and Bichena. With this type of data there is no theoretical distribution to which the analyst can turn to assess the validity of response. It may be that responses from Metekel have been incorrectly coded 1 instead of 7, so the original forms should be checked again. But there is no further analysis to summarise or classify this type of data other than to express the frequencies as a percentage. Presentation of tables is examined in Chapter 12.

hard to interpret. Third, each class must be carefully specified to ensure that every observation belongs to a unique category.

For example, if crop output is classified in 100 kg intervals, the ranges could be specified 0–100, 100–200, 200–300 etc. But in such a case a weight of exactly 200 kg could legitimately be placed in either the 100–200, or 200–300 class. The correct specification depends on the precision of measurement. To an accuracy of 1 kg, classes of 0–99, 100–199, 200–299 etc would be unique. A better definition, to cater for any precision, is 0–<100, 100–<200, 200–<300, etc.

The frequency distributions in Table 9.2 are repeated below in Table 10.1 recalculated at intervals of 5 kg.

The three sets of data illustrate a number of important features. List 1 has the observations clustered in two classes, 5–<10 and 10–<15 kg. We can deduce nothing from the results because they are too bunched. The class intervals are not suitable for the data. Looking back at Table 9.2 a better choice would have been intervals of 2.0 kg or perhaps 2.5 kg. The second and third lists have a better spread amongst the classes. But we can see that there is a central concentration at 10–<15 kg in List 2, whereas the classes in List 3 are more equally distributed.

Note that a separate zero class is included. When dealing with production or performance data it is a good idea always to show zero as a separate class, not included in the first range, such as 0–50 kg. This is for two reasons. Zero production may indicate specific problems amongst the respondents and needs to be identified separately from low production. Secondly, there is a danger that zero values exist in error for missing value codes. Separate identification will help spot this problem. It will also aid calculation of a mean of non-zero observations, as well as an overall mean.

Useful though the table is, it is hard to obtain a good feel for the data from the numerical distributions alone. A visual presentation would help understand the shape, and some summary statistics are needed to describe it.

Visual presentation

Histograms drawn from Table 10.1 are shown in Example 10.2 . The characteristics of each list are immediately visible: a bunched peak for List 1, a distinct pattern for List 2 and a shapeless, almost uniform distribution, for List 3. The reason for looking at the distribution is to determine how regular the shape is, the extent of any central tendency in the data, and the existence of extreme high or low values. Extreme values

Table 10.1 *Frequency distributions of unthreshed yield*

Yield class (kg/100 sq m)	List 1	List 2	List 3
0	0	1	1
>0–<5	1	0	5
5–<10	12	5	4
10–<15	21	12	5
15–<20	0	5	6
20–<25	0	3	5
25–<30	0	0	2
30–<35	0	1	2
35–<40	0	1	0
Total	34	28	30

EXAMPLE 10.2

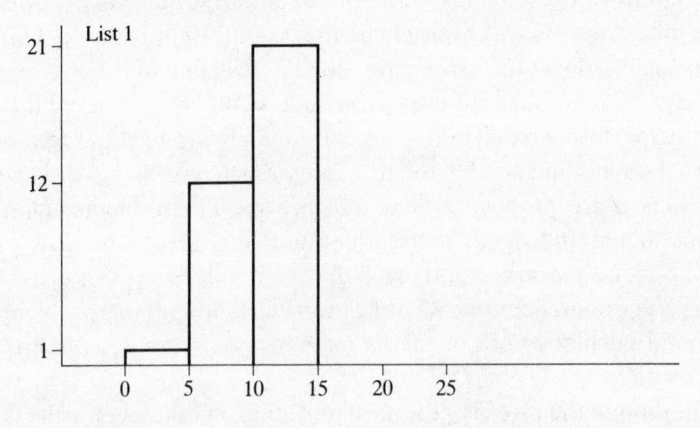
List 1
21
12
1
0
5
10
15
20
25

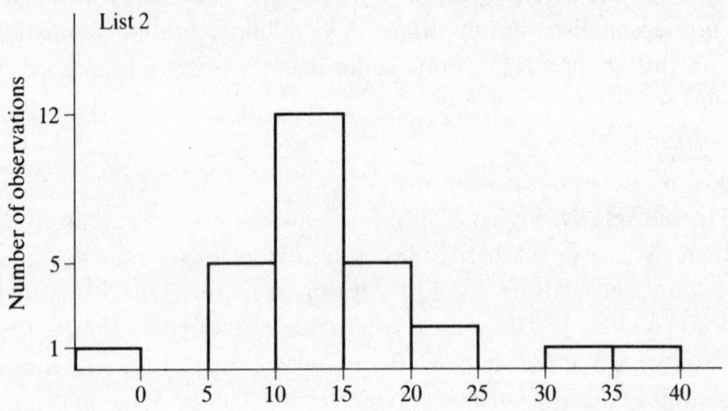
List 2
Number of observations
12
5
1
0
5
10
15
20
25
30
35
40

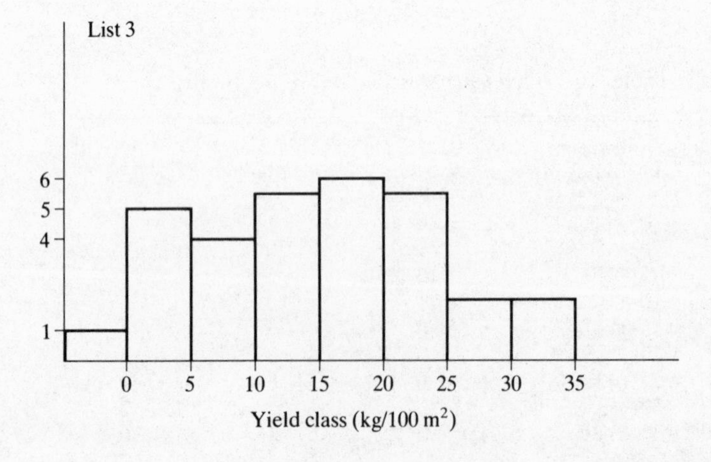
List 3
6
5
4
1
0
5
10
15
20
25
30
35
Yield class (kg/100 m²)

are referred to as outliers, because they lie outside the main concentration of observations. Outliers may be valid observations, but in view of their potential effect on the mean and standard deviation they should be identified before the survey results are interpreted, and checks made back to the original survey documents, or even to the field, if necessary.

Frequency distributions by class intervals allow a quick inspection of the range of data. But they would not reveal the data fabrication problems described earlier, in Chapter 9, because details of individual observations are lost. A useful exploratory technique which combines the visual impact of the frequency distribution with individual details is the Stem and Leaf display developed by Tukey and described in his book *Exploratory Data Analysis* (1977). This technique retains the simple ranking of data, but also indicates the shape of the distribution. The three frequency distributions from Table 9.2 are presented as stem and leaf displays in Example 10.3.

EXAMPLE 10.3

Stem and leaf presentation

List 1

3	**4**
7	**3 3 4 4 4**
8	**2 2 3**
9	**1 2 3 4**
10	**1**
11	**0 2 2 2 2 2 3 3 3 3**
12	**1 3 4**
13	**0 1 2 3 3**
14	**3 3**

List 2

0	**00 80 80 80 90 94**
1	**00 00 06 20 20 20 20 30 30 40 40 40 50 70 80 81 90**
2	**00 00 10**
3	**05 80**

(continued)

List 3

0	**00 22 32 34 45 45 52 55 65 71**
1	**12 15 20 21 30 62 70 75 82 91 92**
2	**01 02 04 10 20 51 91**
3	**02 11**

The stem is to the left of the vertical line, and is formed by separating the digits for each value. List 1 has been formed by separating the observations at the decimal point, so the stem has the values of whole kilogrammes. The decimal values are listed to the right. Thus the first observation, 3.4 appears as 3 to the left of the bar and 4 to the right. Every observation is shown, so repetitions are immediately evident. Lists 2 and 3 have the stem divided into ten kilogram classes. Thus the first observation in list 2 appears as 0 stem and 00 leaf, then 0 stem and 80 leaf for 8.0 kg. The last in List 2 is set out as 3 stem and 80 leaf for 38.0 kg. Work through the data from Table 9.2 until you understand the method. The choice of break point for the stem will obviously influence the layout and the visual impression gained from the physical arrangement of the numbers. The visual impression from the stem and leaf differs from the histogram, even though both are the same data, because of the scale and break points chosen. One disadvantage compared with a simple frequency diagram is that the data have to be ranked, to display repetitive sequences and convey the full effect of the layout.

The data in the stem and leaf were introduced in Chapter 9 to illustrate fabrication. List 1 contains multiple cases of the values 11.2 and 11.3; List 2 has a preponderence of values accurate to a whole kilogram; and List 3 has a limited range of decimal values. With the break points as chosen, the characteristics of Lists 1 and 2 show up clearly in a way that is impossible with a frequency distribution. The use of a stem and leaf would alert the analyst to a data problem before the start of summarising the results. The problems with List 3 are not so evident. A different break point, such as the decimal place, might reveal the problem, but the range of observations would lead to a long stem with few leaf values and therefore a weaker visual impact.

The shape of the histograms and stem and leaf distributions can be compared with the normal distribution and variations of the normal distribution to identify skewness and kurtosis (how peaked or flat the shape is). Example 10.4 shows unimodal (single peak) distributions which are symmetrical, skewed, flat and peaked. Data from smallholder

EXAMPLE 10.4

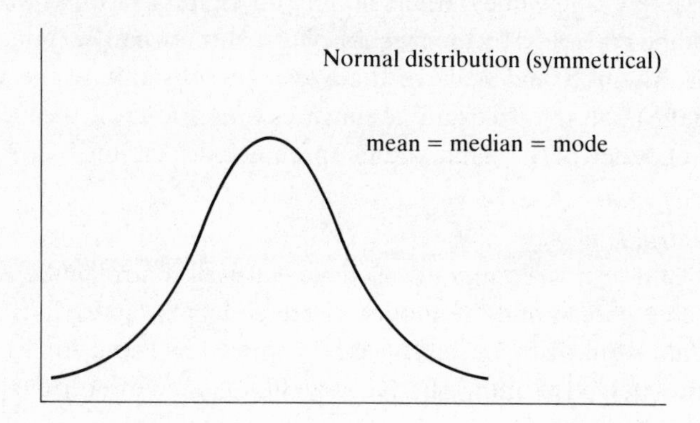
Normal distribution (symmetrical)
mean = median = mode

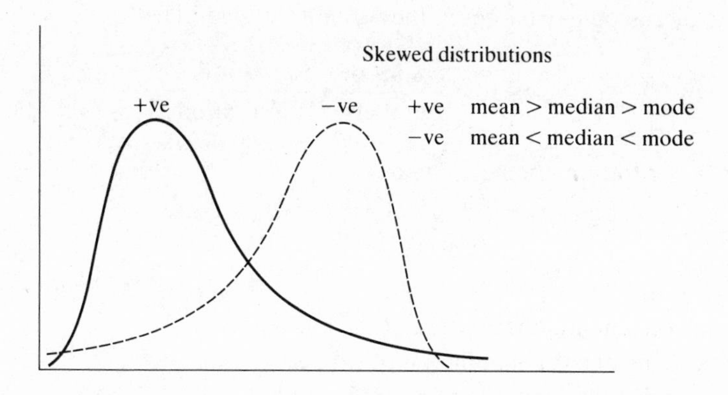
Skewed distributions
+ve
−ve
+ve mean > median > mode
−ve mean < median < mode

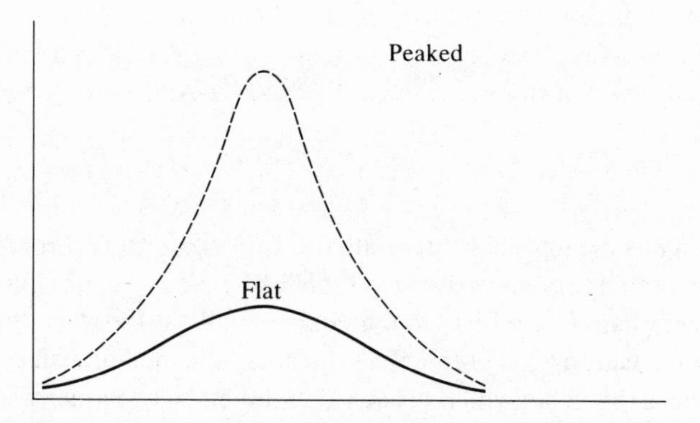
Peaked
Flat

farmers are typically skewed with a concentration of low value observations and a few high value outliers in the long right tail. But as we have seen, the shape conveyed by the frequency distribution can be influenced by the class intervals and scale of the axes. To substantiate the visual impression the analyst must examine statistics which describe the location or central tendency of the data, and its spread or dispersion.

Central tendency

The three most commonly used measures of central tendency are the mean, the median and the mode. These concepts can be illustrated using the data from List 1 in Table 9.2 reproduced in Table 10.2. Recall that the data referred to unthreshed crop yield obtained by crop cuts from 100 square metre subplots.

As discussed in Chapter 2, the *mean* is defined in words as

$$\text{mean} = \frac{\text{sum of observations}}{\text{number of observations}}$$

This is written mathematically as

$$\bar{x} = \frac{\Sigma x_i}{n}$$

where x_i is the ith observation of the variable x,
Σ means the summation of values,
n is the number in the sample, and
$\bar{x}$ is the mean.

From List 1,

$$\bar{x} = \frac{355.50}{34} = 10.46$$

The *median* is the middle value when the data observations are ranked in ascending or descending order. Thus, 50% of observations lie on either side of the median. From List 1 the median is halfway between the 17th and 18th observations. As both are 11.2, that is the median value.

The *mode* is the value which occurs with the highest frequency. From List 1 this is 11.2. With continuous data, the data are normally first grouped into a frequency distribution. The mode is entirely dependent on the classes chosen for the distribution so comparison is restricted to other distributions with identical class intervals.

If the distribution is symmetrical the mean is the preferred measure because it is calculated from all the observations and is statistically the most precise. With a skewed distribution the median is preferred as it is less influenced by skewness. The mode is not very useful for continuous or discrete data as it is not unique (because it is dependent on the interval

Table 10.2 *List 1 – Unthreshed crop yield (kg/100 m²)*

		Value (x)	Value x^2
		3.4	11.56
		7.3	53.29
		7.3	53.29
		7.4	54.76
		7.4	54.76
		7.4	54.76
		8.2	67.24
		8.2	67.24
Lower quartile –		8.3	68.89
		9.1	82.81
		9.2	84.64
		9.3	86.49
		9.4	88.36
		10.1	102.01
		11.0	121.00
Mode		11.2	125.44
Mode	Median –	11.2	125.44
Mode		11.2	125.44
Mode		11.2	125.44
Mode		11.2	125.44
		11.3	127.69
		11.3	127.69
		11.3	127.69
		11.3	127.69
		12.1	146.41
Upper quartile –		12.3	151.29
		12.4	153.76
		13.0	169.00
		13.1	171.61
		13.2	174.24
		13.3	176.89
		13.3	176.89
		14.3	204.49
		14.3	204.49
Sum Σ		355.50	3918.13
		n = 34	

chosen for a frequency distribution) and in the case of a uniform distribution may not exist at all. It is only useful for nominal data (see Example 10.1, where the mode is Category 5, Bichena district).

Dispersion

The average tells us where the data are centred, but nothing about their variation.

One way to quantify variation either side of the mean would be to add up the differences between the mean and each observation. But as the mean is the central point, the sum of simple deviations will be zero. Instead, we can calculate the sum of the square of those differences. After dividing by the number of observations we obtain the mean square deviation – or the variance. The standard deviation is the square root of the variance.

$$\text{Variance} = \frac{\Sigma\,(x_i - \bar{x})^2}{n}$$

$$\text{Standard deviation (SD)} = \sqrt{\text{Variance}}$$

The formula can be manipulated algebraically so that it is simpler for computation. We now also use $n - 1$ instead of n for the denominator as it gives better *estimates* of the standard deviation of the *population*. It *must* be used for small samples where $n < 30$, and with large samples the difference, whether using n, or $n - 1$ is negligible.

$$\text{Standard deviation} = \sqrt{\left(\frac{\Sigma x^2 - \dfrac{(\Sigma x)^2}{n}}{n - 1}\right)}$$

This formula requires us to calculate the sum of x and the sum of x^2. Knowing the sample size, the standard deviation can quickly be calculated.

Returning to our data from List 1, we can substitute values into the formula

$$\text{Standard deviation} = \sqrt{\left(\frac{3918.13 - [(355.50)^2/34]}{33}\right)}$$

$$= \sqrt{\frac{201.06}{33}} = \sqrt{6.09}$$

$$= 2.47$$

By taking the square root of the variance, the measure returns to the same units and scale as the mean. A useful comparison then is the standard deviation expressed as a proportion or percentage of the mean. This is known as the *coefficient of variation*.

$$\text{Coefficient of variation,} \quad \text{CV} = \frac{\text{SD}}{\text{Mean}}$$

$$= \frac{2.47}{10.46}$$

$$= 0.24 \text{ or } 24\%$$

As observed earlier, the standard deviation is associated with the mean and is best when the distribution of data is symmetrical. An alternative measure of dispersion follows from the median by dividing the distribution into quartiles.

The *lower quartile* is the value below which one quarter of observations lie. The *upper quartile* is the value above which one quarter of observations lie. The *inter-quartile range* is therefore

upper quartile – lower quartile.

From the data for List 1

Lower quartile:	8.3
Upper quartile:	12.3
Inter-quartile range:	4.0

The lower and upper quartiles describe the distribution of observations either side of the median and are especially useful for skewed data because they reflect the actual distribution.

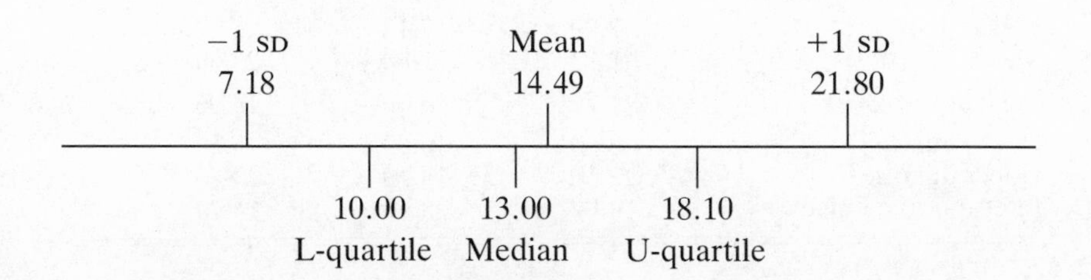

The mean, median, standard deviation and quartiles are calculated for the three lists of data from Table 9.2 and shown in Table 10.3.

The standard deviation is by calculation symmetrical about the mean. For the data in List 2, which are clearly positively skewed (see Example 10.2), the range of standard deviations about the mean describes the shape of the data less accurately than the inter-quartile range.

An important characteristic of these descriptive statistics is their stability. Although the mean and standard deviation are the most often quoted statistics they are in fact very susceptible to extreme values. For example, the mean and standard deviation of a series of fifteen numbers 1,2,2,3,3,3,4,4,4,4,5,5,5,5,5 are 3.67 and 1.29. If a single high value number such as 25 is added to the series the mean becomes 5, and the standard deviation 5.48. By comparison, the median for the original series was 4 and for the expanded series is still 4. The mean was increased by 36% and the standard deviation went up 325%. This sensitivity of the standard deviation is because it is calculated from squared deviations about the mean. The mean is responsive to outliers, but extreme values at one tail of the distribution can be offset by extreme values at the other tail, whereas the standard deviation increases with extreme values at either tail because the deviations are squared.

The sensitivity of these statistics means that the standard deviation will give a poor description of the spread of the data if the distribution is skewed and contains outliers. So data should always be checked for outliers before relying on the standard deviation. Similarly, the median is a better estimate of central tendency than the mean, in conditions where isolated extreme observations enter the data set. However, the sensitive

Table 10.3. *Descriptive statistics*

	List 1	List 2	List 3
Mean	10.46	14.49	14.29
Standard deviation	2.47	7.31	8.81
Coefficient of variation	0.24	0.50	0.62
Mean −1 SD	7.99	7.18	5.47
Mean +1 SD	12.93	21.80	23.09
Median	11.20	13.00	13.00
Lower quartile	8.30	10.00	5.50
Upper quartile	12.30	18.10	20.20
Inter-quartile range	4.00	8.10	14.70

property of the standard deviation can also be used to advantage. A high coefficient of variation can be interpreted as an indicator of extreme values, and the need for further investigation from the survey forms or even a field visit.

A large standard deviation in relation to the mean is undesirable. It means the data are very variable and would therefore have a large confidence interval. If a large value is due to outliers which are caused by errors in data collection or processing the apparent variation will not be a true representation of the population. Because of this uncertainty with the standard deviation it is good practice always to present *both* a frequency distribution *and* the mean and standard deviation when reporting results. In this way the shape of the distribution can be compared with the calculated statistics.

Although the median is a valuable statistic because of its stability, calculation requires the data set to be ordered. Ordering is straightforward on a computer, but time-consuming by hand, especially if repeated calculations are wanted for subpopulations, and as a result the median is seen less frequently than the mean. For nominal or ordinal data, the mode is the only relevant measure of central tendency.

Classifying data for two variables

In a similar manner as for single variables, the first step to exploring the data will often be a joint frequency distribution. The principles are the same as for individual data of nominal or ratio measurement.

In this case two ratio variables, one continuous and the other discrete are tabulated against each other to identify the relationship between

Table 10.4. *Joint frequency distribution – Farm size by oxen ownership*

	Number of oxen			
Farm size (ha)	0	1	2	3+
<1.0	15	11	2	0
1.0–<2.0	14	20	9	1
2.0–<3.0	2	12	15	6
3.0–<4.0	0	12	18	10
≥4.0	1	3	10	5

oxen ownership and farm size. In the case of two continuous variables it may not be possible to specify the appropriate classes for a distribution without first inspecting the data. The best approach is to plot a scatter diagram (Example 10.5). Each variable is assigned to one axis of a graph. The values of each observation make up the coordinates of its location on the plot. Thus in a plot of farm income (Y axis) against farm size (X axis) the income derived from the farm represents the distance along the Y axis, and the size of farm the distance along the X axis. The existence of a relationship between the data may be inferred from the pattern of coordinates marked on the plot and appropriate statistical tests of relationship made. The scatter diagram enables a speedy visual check of outlying observations, as well as clusters or other patterns in the data.

Descriptive statistics and sample estimators

In the introduction to this section it was pointed out that calculations of population estimates from survey data must be made according to the type of sample design used. The reader may be puzzled about the relationship between the simple descriptive statistics discussed

---

EXAMPLE 10.5

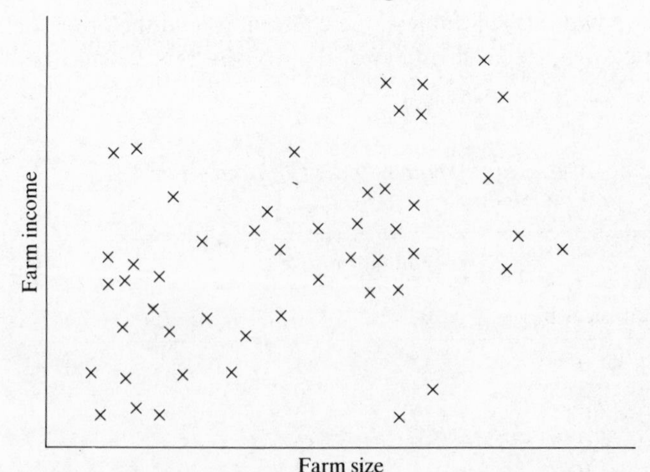

---

earlier and the procedures for estimating totals, means and ratios introduced here and in Appendix 1. Confusion tends to arise about the circumstances in which it is correct to use descriptive sample statistics to estimate population values and about the terminology of variance, standard deviation and standard error.

Descriptive statistics are used to summarise data in order to identify the point(s) around which the data are located (central tendency), and the spread or dispersion of the data. With a self-weighting design, where every unit has an equal chance of selection, a simple arithmetic mean calculated from the sample will equal the population mean calculated by the methods shown in Appendix 1. If the design is not self-weighting, some observations will over- or under-represent the subpopulation from which they are selected compared with the rest of the sample. In this case a simple arithmetic mean would not be equal to the correct population calculation, giving rise to a bias in estimation of the true mean.

A similar argument applies to the simple formula for standard deviation. If all observations are collected with an equal chance of selection a sample frequency distribution will correctly represent the population, and the standard deviation will describe the distribution of those data. If the data contain a mix of observations with different probabilities of selection, unweighted calculations of variance and frequency distributions would mis-represent different subpopulations because they are over- or under-represented in the sample compared with the population. When the data have been sampled at a constant probability the simple descriptive statistics will give a true representation of the population. Subpopulations sampled with different probabilities should either be weighted or analysed separately for descriptive statistics.

A different issue is the variance and its square root. The simple variance calculated from the data is the variance of the data about the mean, and its square root is the standard deviation. Variances described in Appendix 1 are variances of the estimates, not of the data, and their square roots are standard errors (standard deviations of sample estimates as described in Chapter 2).

The standard error can be calculated from the standard deviation of the data from a simple random sample in the following way as demonstrated in Chapter 2.

$$\text{Standard error} = \sqrt{\frac{\text{Variance}}{\text{Sample size}}}$$

This calculation holds true only for a simple random sample. The standard error of an estimate is a characteristic of the sampling distribution. The simple statistics described earlier can be used to present the data, but calculations involving the use of standard errors, such as for tests of differences of means, totals and ratios between subpopulations, and the setting of confidence intervals, should use the correct calculation of standard error for that particular sample design.

As a general principle use the simple arithmetic mean and standard deviation to explore the data, but use the correct estimators as described in Appendix 1, when presenting population statistics and sampling error.

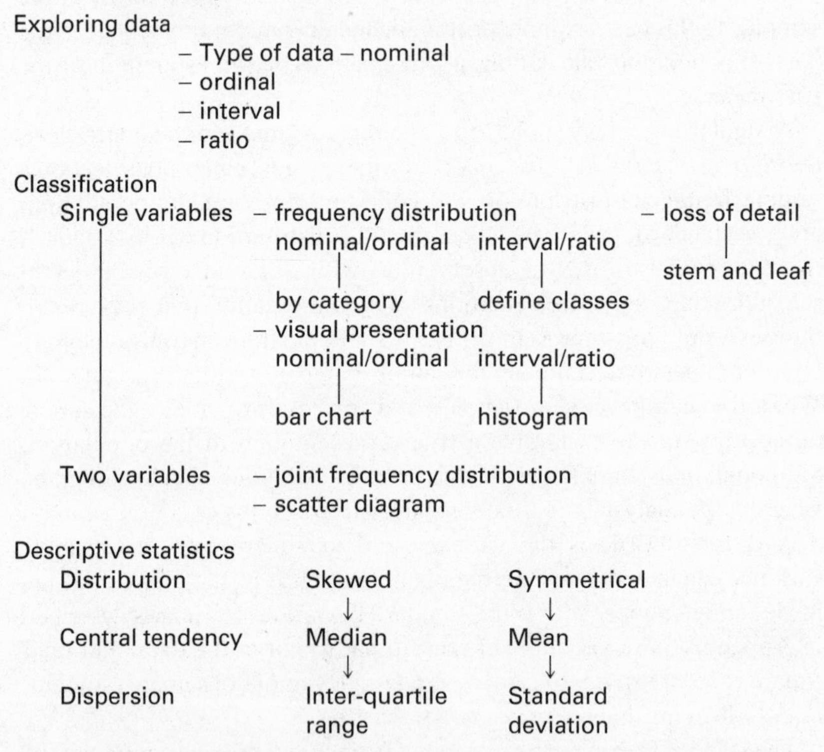

10.2 Estimation from survey data

Population estimates from a simple random sample

Once a preliminary exploration of the data has been made, the next step, before turning to measures of comparison, will be to calculate estimates for population values. Even where the main purpose of the survey is to investigate and make comparisons between subpopulations, it will often be necessary to prepare summary tables of population totals.

Calculations of population estimates are derived from the type of sampling design used for the survey. If the design gives an equal chance of selection to all units in the population, the calculation of population estimates is straightforward. Calculation of variances, and the estimation of population values for samples where unequal selection probabilities were used, requires more complex calculations.

In this chapter we explain the procedure for a simple random sample, and refer to examples from other designs in Appendix 1.

Because a simple random sample (SRS) has a simple structure, with equal probabilities of selection and a single stage, the formulae for estimating characteristics of the population are also straightforward. From our survey data, we usually calculate four types of estimate:

- an average or mean value, such as family size
- a total value, such as crop area or number of cows in a population
- a ratio, such as crop production per unit area (yield per hectare)
- a proportion, such as the number of cattle dipped last month as a proportion of all cattle

To introduce the formulae for estimation we must first introduce the notation. The letters Y and X are used to denote variables, such as the area of sorghum per household, or the number of oxen owned by a farmer. Capital letters indicate the value in the population (which is usually not known) and lower-case letters indicate the value in the sample. Thus Y could be the population area of rice, and y the area measured in the sample. The letters N and n generally indicate the number of units of study in the population and the number in the sample, respectively. The subscript $_i$ is used to indicate individual observations in a sample. Thus from a sample of five plots the five areas would be denoted as y_1, y_2, y_3, y_4, y_5. Each value of y indicates one of the five observations. The term y_i refers to any of the sample observations for the variable Y.

The mean of a variable is denoted by a bar placed over the symbol for the variable. Thus the sample mean for the variable x is $\bar{x}$, and the population mean is $\bar{X}$.

A population value which has been estimated from the sample is indicated by a cap ($\hat{}$) above the population symbol. Thus, estimated population totals and means are indicated as follows: total $\hat{Y}$, mean $\hat{\bar{Y}}$.

Turning now to the formulae, if we consider a variable Y which has values y_i for the n units in the sample, then we have:

$$\text{Population mean } \hat{\bar{Y}} \text{ estimated by } \frac{\Sigma y_i}{n} = \bar{y} \text{ the sample mean} \quad (10.1)$$

$$\text{Population total } \hat{Y} \text{estimated by } \frac{N}{n}\sum y_i = N\bar{y} = \frac{1}{f}\sum y_i \quad (10.2)$$

where N is the population size, and $f = n/N$, the ratio of the sample size to the population size, is known as the sampling fraction.

The estimation of ratios and proportions is as follows where X is a second variable:

$$\text{Population ratio } \hat{R} = \frac{Y}{X} \text{ estimated by } \frac{\bar{y}}{\bar{x}} = \frac{\Sigma y_i}{\Sigma x_i} \quad (10.3)$$

$$\text{Population proportion } \hat{P} = \frac{A}{N} \text{ estimated by } \frac{a}{n} = p \quad (10.4)$$

where A is the number of units in the population falling in the relevant category, and a is the number in the sample.

Having calculated the population estimates we now want to know how close these estimates can be expected to be to the mean of their sampling distribution. In other words we want to know the standard errors of the estimates so as to have a measure of sampling variability. (See Chapter 2.) For the SRS design, the formulae for estimating the variances (the squares of the standard errors) of the estimates are given below. The variance of an estimator is indicated by the letter v followed by the estimate in parenthesis.[2]

$$\text{Mean} \qquad v(\hat{\bar{Y}}) = \frac{1}{n(n-1)}(\Sigma y_i^2 - n\bar{y}^2) \quad (10.5)$$

$$\text{Total} \qquad v(\hat{Y}) = N^2 v(\hat{\bar{Y}}) = \frac{N^2}{n(n-1)}(\Sigma y_i^2 - n\bar{y}^2) \quad (10.6)$$

$$\text{Ratio} \qquad v(\hat{R}) = \frac{1}{\bar{X}^2 n(n-1)}(\Sigma y_i^2 + \hat{R}^2 \Sigma x_i^2 - 2\hat{R}\Sigma y_i x_i) \quad (10.7)$$

approximately, provided that n is large.

$$\text{Proportion} \qquad v(\hat{P}) = \frac{p(1-p)}{n-1} \quad (10.8)$$

Throughout, Σ indicates summation, and ^ indicates an estimate. In all cases, the standard error of the estimate is given by the square root of the variance.

Some of these formulae may look complicated and intimidating, but actually they are straightforward to apply in practice and the most complicated calculation is working out either the squares of a column of numbers, or the products of the numbers in two columns, and summing the column. For example, take equation 10.6, the variance of a total. The equation is

Total $$v(\hat{Y}) = N^2 v(\hat{\bar{Y}}) = \frac{N^2}{n(n-1)} (\Sigma y_i^2 - n\bar{y}^2) \quad (10.6)$$

There are two parts to this formula.

$$\frac{N^2}{n(n-1)}$$

This first term consists of the square of the number of units in the population (N) divided by the term $n(n-1)$ where n is the number of units in the sample.

$$(\Sigma y_i^2 - n\bar{y}^2)$$

The second term is similar to a simple calculation of variance. On the left is the sum of each value of y squared, from which is subtracted n times the square of the sample mean.

After this term has been calculated the results from the two terms are multiplied together to obtain the variance. The square root of the variance is the standard error of the estimate. Examples of the calculations for this and the designs discussed in Chapter 3 are given in Appendix 1.

10.3 Calculation of crop areas and yields

Estimation of crop production is often the main purpose behind rural surveys, with a requirement to compare production under different systems, or by location or some technical factor such as a new seed variety. Production statistics are commonplace in national and international compendia, and in the yearly analyses of departments, regions and projects. Those statistics are used to justify extension programmes, research studies and project investment. Yet for such important variables there is considerable misunderstanding about how to report or analyse such data, and in the case of crop yield even how to calculate yield correctly. Because of the fundamental importance of these statistics they are considered as a separate topic here.

Analysis of crop areas

Cropping patterns found in peasant agriculture vary from simple sole crops, such as rice, which are rarely mixed with other crops, to complex intermixtures, often found under low rainfall conditions and designed to maximise returns from unreliable moisture. The more complex these mixtures are, the greater the problem for reporting survey results.

For mixed cropping we refer to the definition used for the 1980 World Census of Agriculture in which a mixed crop is two or more different temporary or permanent crops (but not both temporary and permanent crops) grown simultaneously on the same field. More specific terms such as alley cropping, interplanting, or indigenous descriptions such as Gicci, in Northern Nigeria, may be used to categorise the spatial and sequential structure of the mixture. A situation where both temporary and permanent crops are grown on the same plot (maize, alley cropped with leucaena) is referred to as an associated crop, or if more than one temporary crop is grown, a mixture associated with the permanent crop.

For reporting, there are three different states to be considered. First are sole crops. These are straightforward. Plots sown to the crop are indicated on the survey forms. The area cultivated by each sample unit is summed to obtain estimates of total and average area and their standard errors.

The second group are common mixtures. By common is meant a situation where the mix of crops is planted and harvested at more or less regular times by farmers throughout the survey area and where the ratio of crops in the mixture follows well established norms for the locality. If these conditions exist, the mixture will be found throughout the survey area and can be treated in the same way as a sole crop for the purpose of analysis. Information on both the time of planting and the ratio of crops in the mixture is important to ensure that plot observations, which may be used for analysis of crop response to inputs, are being compared across consistent conditions.

The third situation is where farmers adopt personal strategies of crop mixtures so that over the survey area a large number of different mixtures are reported. In many cases the mixtures will be based around major grain or root crops, but the variation in plant density, spatial arrangement and timing renders comparison between plots difficult. Table 10.5 illustrates a four-crop sequence from West Africa.

During one season the field has seven different mixture states. Many surveys would involve the enumerator subdividing the field into plots containing the same set of crops. But in this illustration full development is not reached until June. Moreover, the mixtures which would be used to categorise the plots in June last at most for two months, yet the plot/crop mixture will be used to describe the land for the whole season.

If the sequence is widespread in the survey area it can be treated the same as a sole crop, but the variations indicated in Table 10.5 are more likely to reflect individual farmer preference. A number of courses is open to the analyst.

Table 10.5. *Sequential planting and harvesting of crop mixtures*

	Crop planted/harvested	Plot crop mixture
March	Melon – on whole field	1 Melon
April	Maize – on half field	1 Melon 2 Maize/melon
May	Sorghum – whole field	1 Melon/sorghum 2 Maize/melon/sorghum
June	Groundnut – part field	1 Melon/sorghum 2 Melon/sorghum/groundnut 3 Maize/melon/sorghum
July	Harvest melon	1 Sorghum 2 Sorghum/groundnut 3 Maize/sorghum
August	Harvest maize	1 Sorghum 2 Sorghum/groundnut
September		1 Sorghum 2 Sorghum/groundnut
October	Harvest groundnut	1 Sorghum
November		1 Sorghum
December	Harvest sorghum	

Reporting crop mixtures

SIMPLIFIED MIXTURES

The first option is to simplify the mixtures by defining plantings at very low densities as scattered plantings. This decision can be left to the enumerator if careful training is given, or made during the coding or data entry period. The aim is to exclude small clumps of crops which could be volunteers, or boundary crops or a little kitchen-garden area not really part of the main field.

SOLE CROP EQUIVALENT

The second approach is to re-allocate the area of the plot on the basis of a sole crop equivalent to the mixture. Different methods have been used based on the number of crops present, seed rate, plant count, yield and value of output. The procedure is straightforward.

$$\text{SCE} = \text{Plot area} * \frac{\text{crop factor}}{\text{sole crop factor}}$$

The sole crop equivalent (SCE) is equal to the plot area multiplied by the ratio of the crop characteristic measured in the mixture to the same measure found in sole crop plantings.

Many problems exist with this calculation, both concerning the choice of a denominator as the sole crop characteristic and the concept of reallocating a mixture back to a sole crop status. For example, consider a 1.2 ha field of millet grown in a mix with sorghum. Two alternative methods to calculate sole crop equivalents are illustrated. The first is based on the number of crops in the mixture, the second on the plant density of the crops.

NUMBER OF CROPS

Using the number of crops the area is allocated according to the number in the mixture. Thus the area of each crop is given as the plot area multiplied by 0.5 for two crops, 0.33 for three crops and 0.25 for four crops. In this way the area of millet plus area of sorghum add up to the plot area but no consideration is given to relative importance. A field of 100 plants of millet and 1000 plants of sorghum would be divided the same as a field with 1000 plants of each crop.

PLANT COUNT

Suppose the plant count for millet in the mixture was 14 250, compared with a sole crop average of 18 500. For sorghum the plant count was 29 500, compared with a sole crop average of 35 000.

$$\text{SCE millet} = 1.2 * \frac{14\,250}{18\,500} = 0.92 \text{ ha}$$

$$\text{SCE sorghum} = 1.2 * \frac{29\,500}{35\,000} = 1.01 \text{ ha}$$

$$\text{Field area} = 0.77 + 0.84 = 1.93 \text{ ha}$$

The calculated SCE areas sum to 1.93 ha which is larger than the actual field (1.2 ha). The SCE can be corrected to sum to the correct area by reapportioning the field according to relative sole crop equivalent area:

$$\text{SCE millet} = 1.2 * \frac{0.92}{1.93} = 0.57$$

$$\text{SCE sorghum} = 1.2 * \frac{1.01}{1.93} = 0.63$$

$$\text{Field area} = 0.57 + 0.63 = 1.20$$

Even though the calculation can be made to work its meaning is questionable. The areas quoted do not actually exist so any statistics of variance or distribution are meaningless.

Redistribution according to an *input* measure ignores any mutual benefits from mixed cropping, which could result in the output from a mixture exceeding the output from two equivalent sole crops (shade, moisture conservation, fertility). This could be taken into account by using yield as the denominator but as with all denominators the difficulty is in deciding which value to use: a survey average from the same year; a national average; a research recommendation? Whichever choice is made will affect the ability to portray the survey area accurately, to make comparison between areas or to make comparisons over time. As a technique it is worth exploring in specific circumstances, but the authors do not recommend it for mixed cropping where the mixtures are very variable.

Multiple tables

The third approach to reporting crop mixtures is the least elegant, but probably the most useful. In order to present the complexity of the cropping pattern several different presentations or views of the data should be given:

1. a table of average or total area sown to each principal crop irrespective of mixture

Table 10.6. *Areas of major crops*

	Mean household Area (ha)	Percentage of households growing (ha)	Survey area (ha)
Sorghum	0.87	87	54 000
Millet	0.21	44	13 000
Melon	0.44	63	27 000
Groundnut	0.21	40	13 000
Yam	0.19	57	12 000
Rice	0.21	58	13 000
All Crops	1.94	–	120 000

2. a table of the proportion of area of each crop grown sole or as a 'number' or named mixture, and
3. a table of area under the principal mixtures

Table 10.6 gives an overview of the main crop areas. Because it will involve double counting of areas under mixtures, and because not all households will grow all crops it is best to include the proportion of growers and total cultivated area in the survey. Note that the total areas are less than the sum of crop areas, due to double counting.

The table gives a ready indication of the relative importance of each crop to the survey area but care must be taken if any of the crops are grown in mixtures as the table will involve double counting. An area under sorghum and millet would be counted once for sorghum and once for millet.

Table 10.7 takes the analysis one step further by considering the distribution of the area of each crop. It should be shown either as principal crops by number of accompanying crops in the mixture, or with named accompanying crops. This will depend on the number and complexity of mixtures.

The first part of the table illustrates the structure of the crop by mixture, but still involves double counting of areas (a mixture of sorghum with millet would be shown under Sorghum – 2 crop mixture *and* Millet – 2 crop mixture). The second part starts to overcome double counting by naming the crop in the mixture. Thus we know that the area of maize with millet would be the same as the area (not shown here) of millet with maize. The percentages in this table can be used with the areas quoted in Table 10.8, which indicates the areas grown by specific mixtures, to calculate actual areas.

Table 10.8 contains the full level of detail, but must be used cautiously. In areas of complex mixtures a very large number of enterprises may be involved. In a survey in Niger State of Nigeria in 1980, a full specification of crop enterprises with no more than 5% of the area allocated to an 'Other' category required over 60 mixture categories. With so many different mixtures, each mixture will be based on only a few observations, unless the sample size was very large. This makes it unreliable for reporting averages of area or density and production. In this situation the analyst must indicate clearly how widespread the occurence of each mixture is.

The thrust of this argument is that it is wrong to try for one statistic or table which will summarise the complexity of crop area. Sequential planting and harvesting of crop mixtures are difficult to quantify and summarise. The examples used here are for a climate with defined

Table 10.7. *Percentage area of crop as:*

	Sole	2 crop mixture	3 crop mixture	4+ crop mixture
Sorghum	32	56	11	1
Millet	21	39	34	3
Yam	84	13	3	0

or alternatively

	Sole	With millet	With beans	With groundnut	Other mixtures
Maize	24	15	32	19	10

Table 10.8. *Area under named crop mixtures*

	Mean household area (ha)
Yam sole crop	0.23
Yam/maize	0.14
Yam/sorghum	0.02
Yam/cassava	0.03
Yam/cowpea	0.04
Yam/maize/cassava	0.03
Yam/sorghum/cassava	0.02
Yam/maize/sorghum	0.02
Yam/maize/cowpea	0.02

growing seasons. Situations exist where there is scarcely any period when crops are not growing, and where groups of mixtures may be scattered in small clumps throughout a field. The most sensible approach in such complicated circumstances is to make use of case study illustrations of specific plots and quantify crop areas by broad mixture categories only.

Analysis of crop yield

Calculating mean yield

Practical aspects of yield data collection have been reviewed in Chapter 5. Our concern here is with calculating yield. The term yield is defined as output per unit area. Examples of ratio calculations for different statistical designs are given in Appendix 1. Yield is described as a ratio estimate: the ratio of output over area. (This is true even if yields are collected by crop cutting, which involves harvesting a known area.) At the level of individual plots the ratio is self evident. But it is common to find calculations of average yield from a survey based on the simple arithmetic mean of plot yields. Except under special circumstances, indicated later, this is wrong, as a simple example will show.

Consider a farmer with five plots of rice (Table 10.9).

What is the mean yield for the farm? The average of the plot case yields is 13931/5 = 2786 kg/ha. But if this is multiplied by the total area of rice (2.48 * 2786 = 6909 kg) we see that the calculation has given a greater output than the household actually achieved. The problem is that different plots have different productivity. Three of the plots have a yield below the mean and two above, but the two which are above the mean are the two smallest, so they contribute a smaller share to total output. The problem with the arithmetic mean is that it treats each plot with equal

Table 10.9. *Plot output and area*

Plot	Output (kg)	Area (ha)	Plot yield (kg/ha)
1	1100	0.45	2444
2	2500	1.25	2000
3	450	0.10	4500
4	1200	0.30	4000
5	375	0.38	987
Total	5625	2.48	13931
Average	–	–	2786

importance or weight. If plot area and crop yield are in any way related then the plot case mean will be biased.

The solution is to calculate the ratio of output to area for the whole farm. The correct yield which equates output with area is Total Output/Total Area: 5625/2.48=2268 kg/ha. This is now the household case yield. If crop yields were collected by crop cutting of subplots, plot output must first be calculated from the subplot yield estimate multiplied by plot area, before summing output and dividing by total area for the farm.

$$\frac{\Sigma\,(\text{plot area} * \text{plot yield})}{\Sigma\,(\text{plot area})} = \frac{\text{Total output}}{\text{Total area}}$$

In this way the yield of each plot is *weighted* by the share of the plot area in total area – in other words by its relative contribution to output.

The correct calculation of yield will depend partly on the sample design and partly on the purpose of the analysis. A basic principle is that survey results should be reported at the level of the unit of study. Thus for a farm survey in which households are the unit of study, statistics such as crop yield and area should be reported at the household level. In order to collect crop data, plots farmed by the household will be used to record information, but the results should be aggregated to the household level for reporting. This applies both to all plots growing a certain crop and to plots with a particular characteristic such as use of an input.

For analysis of household yields the problem of area bias still holds true. Different households will farm different areas and the yield must be weighted by area to correct for any bias. Once more, the correct calculation is the sum of household output divided by the sum of area. Ratio estimators described in Appendix 1 use this approach.

It will sometimes be necessary to sample plots directly, rather than from a household sample, in order to take crop measurements. The correct calculation will depend on the sampling scheme. If the plots are selected with probability proportional to area, a simple arithmetic mean yield can be used because the sample scheme has the effect of weighting each observation according to area. If, as is more likely, a simple random sample is used, yield calculations should be weighted by plot area, so a ratio estimate is required.

The only other circumstances in which a plot case simple arithmetic mean, or a household case simple arithmetic mean, would be correct would be if the purpose is to estimate household or plot performance, for which the condition being studied relates to the behaviour of the farmer or operator rather than the performance of the crops over the survey area. An example would be mean yield of farmers participating in an extension programme for which the analysis is concerned with the farmer's response to extension. But a study of crop yield by use of fertiliser would need to be calculated from output divided by area for all land on which fertiliser was used in case there is a relationship between fertiliser use and plot size.

In order to distinguish between the two calculations the term *yield rate* is advocated for total output divided by total area and *mean yield* for the arithmetic yield of plot or household observations.

Frequency distributions

Because there sometimes exists a relationship between crop yield and plot area it is more informative to present frequency distributions of yield observations as a distribution of the crop area under each yield class, rather than as the number of plot observations or household observations. In this way the frequency distribution indicates the distribution of land by yield class. A frequency distribution of number of plots is potentially misleading because the definition of a plot is usually designed to meet the needs of the survey, rather than because it is of significance for production. Table 10.10 contains a distribution of land area by yield class. The table shows area as a percentage. Total area surveyed is indicated at the foot of each year column, together with yield rate. The table is interpreted as follows. In 1980/81, 35% of the sample area growing yam had yields above 10 tonnes, but this fell to 16% in 1981/82.

Mean, median or output/area

Different methods of yield calculations can give rise to large differences in reported results. Table 10.11 illustrates plot case, household case and area based statistics of median and mean from a survey in Nigeria.

Yield rate is the calculation of output divided by area, which is the unbiased estimate of crop yield. The plot case simple arithmetic mean overestimates yield rate for all six crops, and the household case mean is an overestimate for four. In all but one of the crops the area based median is lower than the yield rate, indicating a skewed distribution. Plot based and household based medians are higher than the area median for all crops except cowpea.

The conclusion to be drawn from this table is that the most likely effect of incorrect yield estimation is to overestimate crop yield. If yield figures are to be used to calculate production from given areas, these overestimates would seriously distort the result.

Summary

Reporting crop areas – sole crops
– constant mixtures | – averages and totals
– variable mixtures
  – sole crop equivalents (but areas meaningless)
  – multiple tables
    – total area
    – area in type of mixture
    – area in specific mixtures

Crop yield — ratio of output to area
– for plots
– for households
– for sample yield average
  [not simple arithmetic plot mean]
– frequency distribution of area by yield class gives best presentation
– incorrect yield calculations tend to overestimate

Table 10.10. *Distribution of crop area by yield category*

	Area (%)	
Yam (kg/ha)	1980/81	1981/82
Zero	0	3
>0–2000	0	2
2001–4000	4	4
4001–6000	18	19
6001–8000	21	35
8001–10000	22	20
10001–12000	8	10
12001–14000	4	1
14001–16000	8	2
16001–18000	9	3
18001–20000	0	*
>20000	6	0
Sample area (ha)	52.37	49.16
Yield rate (kg/ha)	9849	7521

* means >0 and <0.5

Table 10.11. *Calculations of crop yield (kg/ha)*

	Plot case mean	Household case mean	Output/ area ratio	Plot case median	Household case median	Area median
Sorghum	475	411	359	397	391	311
Millet	367	287	291	298	279	262
Cowpea	77	67	63	88	51	92
Groundnut	187	212	146	136	148	118
Maize	441	435	367	419	446	317
Cotton	304	227	250	246	234	179

Source: APMEPU (1982)

Notes and references

1. Chapters 10 and 11 introduce basic principles of statistical analysis: central tendency, variation, and hypothesis testing. Specific references are given for selected topics but the material covered can be found in most introductory texts. The references given in Chapters 2 and 3 are particularly useful: Caswell, 1989; Rees, 1985; Snedecor and Cochran, 1967. More advanced practical guidance is given in Casley and Kumar,1988, Chapter 8.
2. Further readings to explore the mathematical basis of the estimators presented here quickly depart from simple algebra. The authors have based their treatment on Cochran (1977) and refer readers to his Chapters 2 and 3 for background material.

11

Comparative analysis

11.1 Confidence intervals and tests for difference

Setting confidence intervals

Estimates of totals and means are derived from samples, and if repeated samples were taken the results would vary due to sampling error as described in Chapter 2. For each estimate it is important to know the likely extent of sampling error, which is indicated by the size of the standard error. But a more useful presentation for non-technical readers of a report is the confidence interval of the estimate.

The confidence interval is the range either side of the estimate within which the true population value can be expected to lie. It is quoted with a given probability, most commonly 90%, 95% or 99% which represent very high odds of 9:1, 19:1 and 99:1 respectively.

The general principle in constructing a confidence interval for the mean was discussed in Chapter 2. Given that the true population variance was known, this example used the normal distribution to illustrate the construction of confidence intervals (CI). However, it is very common that population variance is not known, in which case an appropriate estimate from the sample will have to be used. In such cases (and particularly for small samples), it is necessary to use the t-distribution: Thus

$$\text{CI for true mean} = \text{estimated mean} \pm t * \text{SE}$$

or:

$$\mu = \bar{x} \pm t * \text{SE}(\bar{x})$$

where t = Student's t value for the degrees of freedom $(n - 1)$ and desired probability, SE = Standard Error of the mean or as calculated for that sample design. A similar principle would apply for construction of confidence intervals for totals, ratios and proportions.

For large sample sizes the value of t approximates z, the normal deviate. In practice, t is used only for small samples where $n < 30$, and when the population variance is unknown – as is often the case.

To take a simple example, assume:

Total area of rice	= 437 ha
Standard error	= 39 ha
Sample size	= 25
From tables at 95% t =	2.06
CI	= 80 ha

The total area of rice is estimated at 437 ha ± 80 ha at 95% probability. The confidence interval is therefore calculated to be 357 ha to 517 ha with a probability of 95%. The converse statement is to say that the chance of the confidence interval not including the population value is just 5% or one in 20. This statement is extremely important for the interpretation of survey results. Too often reports state that an average or total is estimated at an exact figure, but the results from a survey always carry a margin of error. If this error is not set out for the reader a spurious impression of accuracy can arise. In the example above a planner would undoubtedly respond differently if told that the rice area is estimated to be between 357 ha and 517 ha, than if told it is 437 ha. The calculation only takes into account sampling error. As explained in Chapter 2, it is rarely possible to quantify non-sampling error, although it may be as large, or larger than sampling error.

Testing for differences between estimates

Descriptive statements about the survey results are concerned only with estimates and their confidence intervals. But where the purpose is to assess differences between sub-populations, differences between estimates at the same period of time and differences over time, hypothesis testing will be necessary. The most common requirements are to test for differences between means, ratios and frequency distributions.

The reader might pause here to query the need for a statistical test of difference. If one estimate is larger than another then surely there is a difference? There may be, but as we have seen a statistical estimate carries with it a margin of sampling error, described as its standard error. Two estimates which are different could have confidence intervals which overlap, indicating that there is a probability under which the population values could in fact be equal.

Statistical tests of significance take sampling variability into account. A null hypothesis is normally used to frame the situation where no (null) difference exists. In other words that the estimate being tested equals a specified value. The null hypothesis is given the symbol H_0 and can be expressed as:

$$H_0 : \mu = 1500$$

The null hypothesis in this case is that the population mean equals 1500.

The alternative hypothesis H_1 can be expressed as: not equal to, greater than, or less than the stipulated value under the null hypothesis. For example

H_1: $\mu \neq 1500$ Population mean not equal to 1500

H_1: $\mu > 1500$ Population mean greater than 1500

H_1: $\mu < 1500$ Population mean less than 1500

The first example is *two-sided* as the alternative hypothesis would accept a value either greater or less than 1500. The second and third examples are *one-sided* as the only acceptable values are in the second case greater, and in the third case less than 1500.

We know from Chapter 2, that using the properties of the normal distribution, 95% of all sample means should be within 1.96 standard errors of the population mean. Thus if $H_0 : \mu = 1500$ is true then 95% of the sample means should be between 1500–1.96 SE and 1500 + 1.96 SE. This is illustrated in Figure 11.1.

Thus if the value of the sample mean actually falls into *either* area at the extremes of the distribution the null hypothesis can be rejected at the 5% level. By using t or z values, derived from the normal curve, the test can be re-stated in a more useful way. We can calculate the value of t or z from the actual data and compare with appropriate values from the t or z tables.

$$\text{Calculated } t = \frac{\bar{x} - \mu}{\text{SE}}$$

Test: Reject H_0 if $|\text{Calc } t| > \text{Table } t$.

(|Calc t| with vertical bars means the absolute value of t, ignoring the sign of calculated t).

If the alternative hypothesis is that a value is greater or less than a specified value, we need only use half of the distribution.

In a two-tailed test at 95%, the 5% is split between both ends of the distribution – 2.5% at each. In a one tailed test the 5% is located only at one end and therefore it is bounded by a lower z or t value. In this case, illustrated in Figure 11.2, $z = 1.64$ at 95% one-tailed, and $z = 1.96$ at 95% two-tailed.

Figure 11.1. Normal distribution – 95% two tailed.

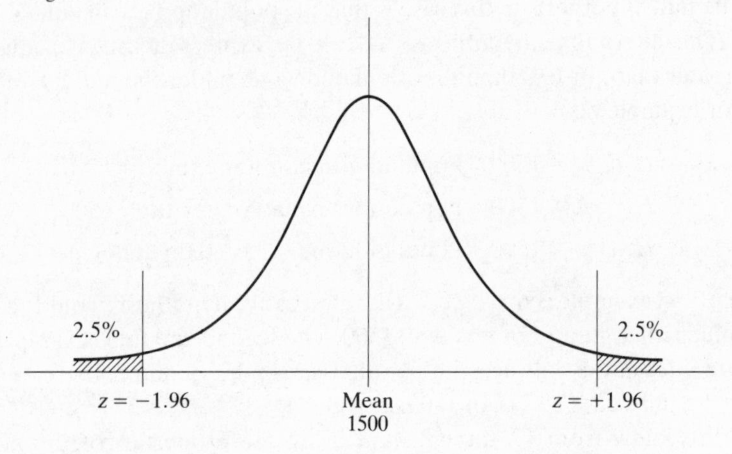

Figure 11.2. Normal distribution – 95% one and two tailed.

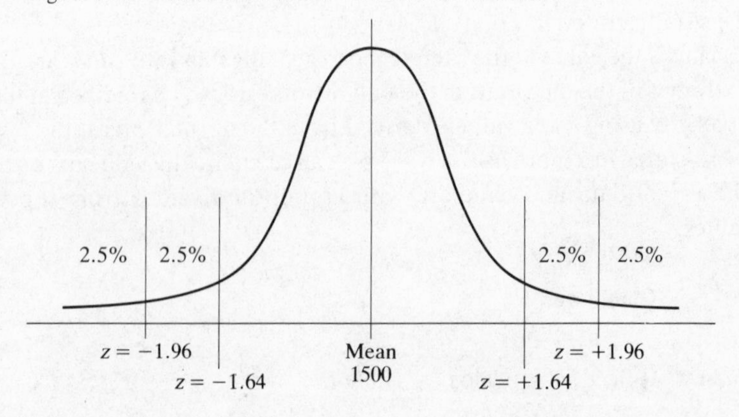

Consider, for example, an experiment to compare the yield of maize from plots treated with a herbicide, with the yield from plots not so treated. The null hypothesis would be that the two average yields are not different from each other. If a test statistic to compare the two averages is significant, the null hypothesis is rejected in favour of the alternative hypothesis that the averages are different. Note that the widely-used term, 'statistically significant' simply means that enough data has been collected to establish that a difference does exist. It does not indicate that the difference is necessarily important. Moreover, a nonsignificant result does not *prove* that the null hypothesis is 'correct' or 'true', merely that it is tenable, that is we do not have sufficient evidence to reject it.

The known probability is chosen according to the degree of confidence which users of the analysis require for the results. If some action is to be taken according to the results of calculated differences a cost will be involved. Statistical tests indicate only a probability for rejecting hypotheses. In the case of the confidence intervals quoted above, although the population value is calculated to lie within a given range at 95% probability there is a 5% chance that it does not. Therefore a decision based on the population value lying within the confidence interval has a 5% chance of being wrong. The confidence interval is linked to hypothesis testing because it defines a range within which any null hypothesis values would be expected to lie (Casley and Kumar, 1988: 84).

Type I and Type II errors

This concept is important for tests of significance. The significance level indicates the decision point for accepting or rejecting the null hypothesis. This determines the probability that a null hypothesis which is in fact true, will be rejected. This is known as a Type I error. The implication here is that incorrect action would be taken in response to the result. For example, consider the assessment of crop yield differences according to herbicide treatment. A Type I error means that a true null hypothesis of no difference in yield according to treatment is incorrectly rejected in favour of a yield difference. A response could be to promote the chemical to farmers in anticipation of a yield benefit. The cost would be incorrect use of the input until a further test was made. The confidence level must be set sufficiently high to minimise the chance of this error.

But equally, there is a second error, Type II, of accepting as true a false null hypothesis. It means, for example, that in a test between two crop yields which shows no statistically significant difference at 95% there is a chance that a difference really does exist. One of the problems of working with sample data is that although most sample estimates tend to be clustered around the population value as we saw in Chapter 2, some are not. Estimates from these samples will not in fact be as close to the true population as the standard error would indicate. A decision not to change a crop extension message based on the calculated difference is in fact wrong. A cost is involved in the delay in promoting a correct message. The significance level must be set sufficiently low to minimise the chance of this error.

Setting the level of probability involves a compromise between committing either error. Raising the significance level increases the chance of a Type II error, lowering it increases the chance of a Type I error. The calculation of the probability of a Type II error depends on the type of estimator being used. Generally speaking, the probability of a Type II error is larger than of a Type I error because the Type I is what the investigator wants to avoid.

If action is to be taken when a significant difference is observed then the Type I error is an error of commission, taking action when it is not justified; and the Type II error is an error of omission, not acting when action should have been taken. The relative costs of both errors are the only guide as to the most suitable probability. But this argument also highlights the need to avoid making decisions on test statistics alone. A significance test should be treated as just one aspect of an assessment of difference. Other factors including data patterns, prior knowledge and observed trends should be considered before a decision is made.

Testing differences for means and ratios

In most practical circumstances the sampling distribution of means and ratios will be a reasonable approximation to the normal distribution. This means that the t and z distributions can be used to test differences. Throughout this section reference is made to the t statistic, which applies to small samples, generally considered less than 30. Except where indicated procedures apply equally to the z statistic for the normal distribution of large samples.

Most often it is desired to test the difference between two means: either from two sub-populations, such as crop area in two regions, or household income by type of occupation; or from the same statistic at two different points in time. The method of testing depends on whether or not the variances of the two samples are from the same population. The technical term for this is that the two sets of data are homoscedastic. This can be tested for, using the F-ratio test between the variances of the two data series containing n_1 and n_2 observations respectively. The test criterion is $F = s_1^2/s_2^2$ where s_1^2 and s_2^2 are the variances of the two samples and s_1^2 is the larger of the two. F has $(n_1 - 1)$ and $(n_2 - 1)$ degrees of freedom.

The null hypothesis is that s_1 and s_2 are from independent random samples from normal populations with the same variance σ^2. The alternative hypothesis is therefore H_1: $\sigma_1^2 > \sigma_2^2$. Calculated F (one-sided) is compared with Table F. If Calculated F is larger, the null hypothesis is rejected. If there is good evidence that the estimates are from the same population then a pooled estimate s_t of the standard error from the two samples can be calculated. Where x_1 and n_1 are the mean and sample size for the first group and x_2, n_2 for the second group, the function $(x_1 - x_2)/s_t$ follows the t distribution with $n_1 + n_2 - 2$ degrees of freedom. If the estimates appear to be from different populations (which may be quite common when working with skewed populations) a modified test is required and the reader is referred to statistics texts (see Snedecor and Cochran, 1967: 114).

t tests of significance

(i) DIFFERENCE FROM A STANDARD VALUE

Using the estimate of household size in Appendix 1, Table A1.4, we wish to test whether the mean household size of 6.0 found in the survey differs from the population census value of 6.5 which had previously been used in the area. A difference in household size would affect the estimate of total population and subsequent calculations about food requirements. The standard error of the estimate was found to be 1.25. Sample size was 26 households.

Thus:

$$t = \frac{\bar{x} - \mu}{\mathrm{SE}(\bar{x})} \tag{11.1}$$

$$t = \frac{6.0 - 6.5}{1.25}$$

$t = 0.4$ (ignoring the sign)

From a table of the t distribution, at a significance level of 5% the region of rejection is given as $t \geq 2.06$ at 25 degrees of freedom. The calculated value is less than 2.06 so does not fall in this region. The null hypothesis is not rejected. The household size in the survey population is not significantly different from the population census figure.

(ii) DIFFERENCE BETWEEN TWO MEANS

Consider the mean household farm income in two areas (Table 11.1).

Table 11.1.

Area	Mean income	Standard error	Sample size
1	365	17	10
2	300	23	15

The standard errors are not significantly different (variances were compared using an F ratio test). As part of the survey analysis we want to see if the apparent difference in income is statistically significant.

For the comparison of the means of two independent samples $\bar{x}_1$ and $\bar{x}_2$ which are estimates of their respective population means μ_1 and μ_2, the t value is calculated from

$$t = \frac{(\bar{x}_1 - \bar{x}_2) - (\mu_1 - \mu_2)}{\text{SE}(\bar{x}_1 - \bar{x}_2)}$$

The denominator is a sample estimate of the standard error of $(\bar{x}_1 - \bar{x}_2)$. It can be shown, but we use without proof, that the variance of a difference is the sum of the variances. Thus for samples with true variances σ^2, estimated by the sample estimates s^2

$$\sigma^2_{\bar{x}_1 - \bar{x}_2} = \sigma^2_{x_1} + \sigma^2_{x_2}$$

with two samples of the same size n this leads to

$$\sigma^2_{\bar{x}_1 - \bar{x}_2} = \frac{2\sigma^2}{n}$$

In most situations σ^2 will not be known, but the best estimate is the pooled average of s_1 and s_2

$$s^2 = \frac{(s_1^2 + s_2^2)}{2}$$

For groups of unequal size, which is more common when analysing survey data, the variance of difference is

$$\sigma^2_{\bar{x}_1-\bar{x}_2} = \frac{\sigma^2}{n_1} + \frac{\sigma^2}{n_2}$$

This still assumes that σ^2 is the same in both populations. If, however, $\sigma_1^2 \neq \sigma_2^2$, the formula still holds

$$\sigma^2_{\bar{x}_1-\bar{x}_2} = \frac{\sigma_1^2}{n_1} + \frac{\sigma_2^2}{n_2}$$

Provided n_1 and n_2 are large SE_1^2 and SE_2^2 can be used to replace σ_1^2/n_1 and σ_2^2/n_2. The t test can then be written

$$t = \frac{\bar{x}_1 - \bar{x}_2}{\sqrt{(\text{SE}_1^2 + \text{SE}_2^2)}} \qquad (11.2)$$

where SE_1 and SE_2 are the standard errors. t is calculated with $n_1 + n_2 - 2$ degrees of freedom.

This is a more useful format as we are dealing with sample data and need to estimate standard errors from samples, using the appropriate formulae for the sample design (see Snedecor and Cochran, 1967: 100–115; Caswell, 1989: 241; Casley and Kumar, 1988: 133).

From the example,

$$t = \frac{365 - 300}{\sqrt{(17^2 + 23^2)}} = \frac{65}{\sqrt{818}} = 2.27$$

From tables the region of rejection at 23 degrees of freedom (10 + 15 – 2, see Table 11.1) is given as $t >= 2.1$ Calculated t is 2.27 which is larger than table t so the null hypothesis is rejected. Household farm income is significantly different between the two areas.

Testing for differences between multiple classes

Testing for differences between two estimates is a frequent form of analysis. It will cater for the most typical survey questions of whether or not the value of a variable differs according to any characteristic for which a simple with or without classification exists: crop yield with or without fertiliser; crop area by traditional or improved variety; income by sex. But many types of analysis involve multiple classifications. The two main ones are: a variable classified into a series of discrete classes such as birth rate by district or family size by ethnic group; and two continuous variables such as crop yield by density of plants, or farm income by area cultivated.

Two different requirements exist here. One is to identify whether or not differences occur between the categories, the other is to identify the magnitude of those differences. For example, when planning a crop extension programme officials may need to know whether to plan separately for different administrative districts or to promote a common package. The first step is to see if the districts differ in cropping pattern or yields. Only if they do will it be necessary to quantify that difference to plan technical advice. A new fertiliser formulation is known to bring a yield response on millet, but to determine the economic efficiency of the product the physical magnitude of response must first be determined in order to calculate economic returns.

Different methods are needed for the discrete categories and the continuous data. Examples of both are given here to illustrate options facing the analyst. All the methods quoted require a full understanding before use. This book does not attempt to cover the methods in depth, for which the reader must refer to one of the statistics texts referenced here and in other chapters.

11.2 Analysis with discrete categories

This section is concerned with three situations:

- analysis of two ordinal variables which are grouped in discrete categories (for example, number of household observations by tractor ownership and participation in extension programmes)
- analysis of one ratio variable which is to be grouped by categories defined by a second ordinal variable (for example, cultivated area by tractor ownership)
- analysis of a ratio variable by categories formed from two ordinal variables (grain production by tractor ownership and participation in extension programmes)

The techniques used are the Chi-square test, and analysis of variance.

To illustrate this approach our example is taken from a survey designed to investigate cropping performance by households according to their use of tractors and their participation in an extension programme. Use of a tractor is classified into ownership, hire, or neither. We wish to know if there is a relationship between tractor ownership and participation and if there are any differences in performance related either to tractor use or to participation. The results will be used to plan a new extension strategy if necessary.

Two ordinal variables

The first step is to set out the response frequencies of tractor ownership and participation in a table (Table 11.2), known as a joint frequency distribution, contingency table or cross-tabulation.

We can see from the column totals that for all categories a minority of households participate in extension. The question is whether this participation is influenced by tractor ownership and use.

In order to test for a relationship we use a Chi-square test. This test examines the probability that the number of observations in each cell of the table would occur by chance. The null hypothesis is that tractor ownership and extension participation (independent from each other) are unrelated and hence the frequency distributions are random across rows or columns. If the distributions are improbable at a chosen level of significance we can reject the null hypothesis in favour of the alternative hypothesis that the two characteristics are related.

The calculation is given by

$$\chi^2 = \Sigma_i \Sigma_j \frac{(O_{ij} - E_{ij})^2}{E_{ij}} \qquad (11.3)$$

where Oij is the number observed, E_{ij} is the number expected in the cell under the null hypothesis. E_{ij} is given by

$$E_{ij} = \frac{L_i * M_j}{N}$$

where L_i and M_j are the marginal row and column totals respectively and N is the overall total.

In Table 11.2, $L_1 = 146$; $L_2 = 487$; $L_3 = 1374$; $M_1 = 551$; $M_2 = 1456$; $N = 2007$

Table 11.2. *Contingency table for households classified by tractor ownership and extension (observed frequencies)*

	Extension participator	Non-participator	Total
Tractor owner	15	131	146
Tractor hirer	152	335	487
Non-owner or hirer	384	990	1374
Total	551	1456	2007

The expected values are calculated in Table 11.3.

The test statistic is now calculated from the sum of observed minus expected squared, divided by expected for each cell.

$$\chi^2 = \frac{(15-40)^2}{40} + \frac{(131-106)^2}{106} + \frac{(152-134)^2}{134} + \frac{(335-353)^2}{353}$$

$$+ \frac{(384-377)^2}{377} + \frac{(990-997)^2}{997}$$

$$= 25$$

The distribution of χ^2 varies according to the degrees of freedom of the distribution given by

$$f = (r-1)(c-1)$$

where r = number of rows, and c = number of columns. In the example

$$f = 2 * 1 = 2$$

First setting the significance level at 5% we consult a table of the distribution of χ^2 with two degrees of freedom and find the region of rejection is where $\chi^2 \geq 5.99$. The calculated value of $\chi^2 = 25$ falls in the region of rejection. We reject the null hypothesis and accept the alternative hypothesis that tractor ownership and extension are related.

Most importantly, it should be noted that the test does not tell us anything about the direction of a relationship or the cause, only that it does or does not exist. In other words we cannot deduce either that

Table 11.3. *Expected frequency table*

	Extension participator	Non-participator	Total
Tractor owner	40	106	146
Tractor hirer	134	353	487
Non-owner or hirer	377	997	1374
Total	551	1456	2007

extension participation *leads* to greater tractor ownership, or that tractor owners or users are more likely to participate in extension.

The χ^2 statistic is not applicable to a contingency table where any cell has an expected frequency of less than 1, or more than 20% of the cells have an expected frequency less than 5. Where these conditions are not met the categories must be grouped so that the expected frequencies are large enough to meet these conditions (Snedecor and Cochran, 1967: 241).

The Chi-square test can be used for larger tables, but a contingency table with only two rows and two columns represents a special case for which a simplified calculation can be used. If the four cells are represented as below

	columns	
	A	B
rows		
	C	D

the calculation is given by

$$\chi^2 = \frac{N(|(AD - BC)| - N/2)^2}{(A + B)(C + D)(A + C)(B + D)} \tag{11.4}$$

where $N = A + B + C + D$ and $|\ |$ represents the absolute value irrespective of sign.

The test should not be used if $N < 20$, or if N lies between 20 and 40, and the smallest expected frequency is less than 5 (Snedecor and Cochran, 1967:221).

The simplified calculation incorporates a correction for continuity because the exact distribution of χ^2 in a 2×2 table is discrete.

An alternative approach to the table for analysis would have been to group tractor owners and hirers together. This would make sense if we had a reason to suspect that the link to extension is due to the use of a tractor, rather than to ownership. If ownership is thought to be the key factor then clearly it must be separated from hire.

Ratio variable grouped into classes

As part of the investigation into farmers who own and use tractors, the next stage of the analysis is to look at the distribution of cultivated area by ownership class. Part of our investigation concerns the extent to which access to cultivable land is concentrated under specific groups of the farming population. One such group is believed to be tractor users, who have the capacity to cultivate a larger area than farmers with animal or hand power. Again, a cross tabulation is a good starting point. There are two approaches to the statistics. We could prepare a table showing the total cultivated area by farmers in each category. This would be valuable for statistics about the survey area, and would enable us to calculate the proportion of the area receiving tractor cultivation. Table 11.4 sets out the total area cultivated by tractor category.

We can see that the largest area is for those farmers who hire a tractor. Calculation of percentages shows that 46% of the cultivated area is by tractor hirers, compared with 19% by owners and 35% by non-owners or hirers. We could now test to see if these totals differ between classes. But we know that the distribution of farmers between classes was not equal. A significant difference could reflect different numbers of farmers rather than difference in cultivated area.

We have already seen that most farmers do not own or hire tractors. Only 146 out of 2007 farmers own tractors. This is just 7%, but they account for 19% of the cultivated area. To examine this we now calculate mean area per household (Table 11.5).

There is a large difference between the average area cultivated per household in each class. Tractor owners cultivate the most (8.6 ha) and non-owners or hirers the least (1.7 ha.). We can conclude from these three tables that

1. tractor owners and hirers are in the minority of the population (146 + 487 out of 2007)
2. but they cultivate the larger share of land (1256 + 3117 ha out of 6709 ha)
3. the difference in area cultivated is due to the higher average area cultivated per household by tractor users.

Table 11.4. *Area cultivated by tractor category*

	Total area cultivated	
	ha	%
Tractor owner	1256	19
Tractor hirer	3117	46
Non-owner or hirer	2336	35
Total	6709	100

Even though the average cultivated area differs between tractor categories, hirers and owners are fairly close compared to non-owners. We should test for significant differences between the categories. We could do three *t* tests: between owners and hirers; owners and non-owners; and hirers and non-owners. But if several *t* tests are performed the probability of making a Type I error increases. The technique to test for differences between more than two means is analysis of variance.

Analysis of variance compares the variation about the mean within a class with the variation between classes. If the ratio of that variation is greater than a test statistic at a predetermined level of significance, the null hypothesis that there is no difference between categories can be rejected. A further calculation can be made to determine if the values for consecutive categories are significantly different from each other. This would enable an estimate of the magnitude of the difference to be made. The calculation is described in standard statistics texts such as Snedecor and Cochran (1967).

There is one important point concerning analysis of variance of cross-sectional survey data. The technique is widely applied to analysis of experimentation in which the number of observations in each category are set equal (a balanced design). This simplifies some calculations. In most surveys the number of observations in each class will not be equal, as in the example here. There are 146 owners, 487 hirers and 1374 non-owners or hirers. Care must be taken, especially if choosing computer software, that the correct formulae are used. Furthermore, the theoretical model includes a number of underlying assumptions concerning the independence and variance of observations. If these do not hold true, results from the test will be meaningless. A thorough understanding is needed before embarking on this analysis.

Table 11.5. *Mean area cultivated by tractor category*

	Mean area cultivated (ha)
Tractor owner	8.6
Tractor hirer	6.4
Non-owner or hirer	1.7
Total	3.34

Analysis of ratio variable by two ordinal variables

The third type of analysis continues from our examples about production by tractor ownership. Having seen that the distribution of tractor use is limited to a minority of the population (146 + 487)/2007 = 32% but that this group cultivate 19% + 46% = 65% of the area, our attention can now turn to the relationship with total production. Cultivated area may be larger for tractor users, but if crop yields are not the same as for hand cultivators their share of total production will be reduced. The first step is a table of total output by tractor ownership (Table 11.6).

The tractor-using group produce 16% + 38% = 54% of production. This is well above their proportion in the population (32%) but surprisingly it is less than their share of cultivated area, which was seen to be 65%. The only explanation for this is if productivity varies between the classes, in other words if crop yield per hectare differs between tractor users and non-users.

The final table (Table 11.7) presents crop yield per hectare by class of tractor use. But because our original concern was with a crop extension programme, and the need to make changes in accordance with the

Table 11.6. *Total output by tractor category*

	Total crop output	
	tonne	%
Tractor owner	2311	16
Tractor hirer	5392	38
Non-owner or hirer	6564	46
Total	14267	100

Table 11.7. *Crop yield (kg/ha) by tractor and extension categories*

	Extension participator	Non-participator
Tractor owner	1753	1850
Tractor hirer	1850	1676
Non-owner or hirer	3125	2688

participation of tractor-using farmers, the table analyses yield by both tractor use and extension participation.

It is immediately apparent that non-owners have a higher rate of crop yield than either class of tractor users. There are also differences between extension participators and non-participators. Amongst the non-owners and hirers participation seems to be associated with higher yields. It just remains to test the significance of these results. Once again, we could make a series of *t* test comparisons, but we would need to do 15 tests to cover all possible combinations. A better approach is a two-way analysis of variance, in which the results are partitioned between both sets of categories. This calculation follows the same principles as for one-way analysis described earlier.

The simple analysis here follows a logical series of steps to look at those aspects of production which could be affected by tractor use and extension participation. We have seen that a relationship exists between tractor use, extension participation and crop production. But we do not know if the yield and area effects are caused by tractors or extension or something else. Other factors could be at work. Perhaps tractor owners have all benefitted from higher education; the social structure within which they live might have favoured allocation of better quality land to wealthy people; extension participators might have been given access to higher yielding seeds, or preferential access to fertiliser. In other words the characteristics we have used to classify the data could in reality be themselves proxies for other factors, or influenced in other ways. The mere existence of a relationship does not mean that one factor causes the other, except under certain restricted experimental conditions. The limitations in analysis are discussed at greater length in the closing section to this chapter.

11.3 Analysis of continuous data

The sort of data referred to here are ratio variables (see Chapter 10) such as income, expenditure, crop area, crop production, etc. The main distinction with the analysis above is that these data form a continuous series and do not have any inherent grouping or classification. The purpose of analysis remains the same: to identify a relationship between two variables and to quantify that relationship.

To illustrate the approach consider a survey in which fields of sorghum were sampled and data collected about the density of plants per hectare and the threshed and dried yield per hectare. The objective of the analysis

Table 11.8. *Plant count and yield data*

	No. of plants (x)/100 m^2	Yield (y) (kg/100 m^2)
1	27	2.4
2	31	4.5
3	51	6.1
4	74	4.2
5	98	3.7
6	105	8.1
7	117	4.4
8	152	5.9
9	156	11.4
10	176	7.7
11	182	7.8
12	195	8.4
13	213	11.8
14	241	13.7
15	248	9.1
16	255	8.8
17	260	10.1
18	315	12.4
19	330	15.0
20	354	8.5

is to test if a relationship exists between sorghum plant density and crop yield.

Twenty observations were collected from subplots measuring 100 square metres, and are listed in Table 11.8 in ascending order of plant population.

Comparison of frequency distributions

During the exploratory analysis, frequency distributions of both plant density and crop yield were calculated. A first step towards quantifying a relationship would be to calculate average yield for each class in the frequency distribution (Table 11.9).

This gives a quick and effective protrayal of average yield and the way in which it varies according to the plant population. There appears to be a steady rise in average yield by number of plants, but with an inconsistent response to higher populations. There are problems with this approach, however. Because the analyst has to classify continuous data into discrete classes, the possibility exists that the choice of classes will affect the interpretation of the relationship. If the data above are recalculated by different classes a new relationship will emerge (Table 11.10).

Table 11.9. *Frequency distribution – 50 plants/100 m^2 class interval*

Plant density classes	Yield (kg)	No. obs.
0–<50	3.45	2
50–<100	4.67	3
100–<150	6.25	2
150–<200	8.24	5
200–<250	11.53	3
250–<300	9.45	2
300–<350	13.70	2
350–<400	8.50	1

The rise in yield by increasing plant population is confirmed, but now a steady response is seen at the higher plant populations.

This example contains few observations and arguably too many classes, but the principle is an important one. The choice of class intervals can lead the analyst, unwittingly, to an incorrect interpretation.

The frequency distribution is a useful summary as long as the limitations of selected class intervals are borne in mind. A better approach for continuous data is a scatter plot, from which a decision can be made to proceed with correlation and regression analysis, if a relationship is thought to exist (see Figure 11.3).

Correlation coefficient

Although the frequency distribution indicates a pattern of response it does not quantify the relationship. To investigate the existence

Table 11.10. *Frequency distribution – 40 plants/100 m^2 class interval*

Plant density classes	Yield (kg)	No. obs.
0–<40	3.45	2
40–<80	5.15	2
80–<120	5.40	3
120–<160	8.65	2
160–<200	7.97	3
200–<240	11.80	1
240–<280	10.43	4
280–<320	12.40	1
320–<360	11.80	2

of a linear relationship between the two variables the correlation coefficient should be used. This statistic has the form:

$$r = \frac{\Sigma xy - \dfrac{\Sigma x \Sigma y}{n}}{\sqrt{\left[\left(\Sigma x^2 - \dfrac{(\Sigma x)^2}{n}\right) * \left(\Sigma y^2 - \dfrac{(\Sigma y)^2}{n}\right)\right]}} \tag{11.5}$$

where x represents the number of plants and y is the yield.

From the data in Table 11.8

$$\begin{aligned} \Sigma xy &= 34\,449 \\ \Sigma x &= 3580 \\ \Sigma y &= 164 \\ \Sigma x^2 &= 824\,670 \\ \Sigma y^2 &= 1575 \\ n &= 20 \end{aligned}$$

Figure 11.3. Scattergram of yield against plant count.

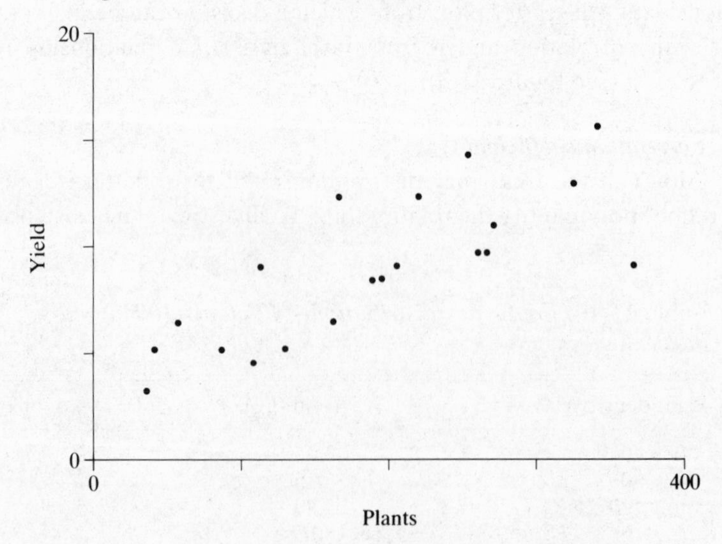

From the formula

$$r = 0.78$$

The correlation coefficient is an indication of the extent to which there is a *linear* relationship between the two variables. It takes the value +1 if there is a perfect positive linear relationship and −1 if there is a perfect negative linear relationship. A value of zero indicates there is no linear relationship (although a non-linear relationship may exist). The value calculated in the example, 0.78 indicates a moderate positive linear relationship. Tables are available to test whether the correlation coefficient estimated from a sample is different from zero, but it is more complicated to test if it is different from a specified non-zero value. The form of the coefficient itself is not linear. A value of 0.8 indicates a stronger relationship than 0.4, but not twice as strong.

Linear regression

Where a relationship is seen to exist between two variables the form of the relationship can be estimated using ordinary least squares regression analysis. In its simple form, this involves expressing one variable (the dependent variable) as a linear function of the second (independent) variable in the following form:

$$Y = \alpha + \beta X + \varepsilon \tag{11.6}$$

where α and β are constants and ε, the error term, is a random variable following a normal distribution with a mean of zero and variance σ^2. α and β are estimated such that the regression line defined by the above equation gives the closest possible fit to the observed values of X and Y. The criterion for that fit is that the sum of squared deviations of observed values Y from the regression line should be minimised (hence known as the method of ordinary least squares). The square of the correlation coefficient between the two variables measures the proportion of the variation in the observed data which is accounted for or explained by the regression line and is called the coefficient of determination.

Using the estimating equations:

$$\beta = \frac{\Sigma xy - \dfrac{\Sigma x \Sigma y}{n}}{\Sigma x^2 - \dfrac{(\Sigma x)^2}{n}} \tag{11.7}$$

$$\alpha = \bar{y} - b\bar{x} \tag{11.8}$$

From the data in Table 11.8

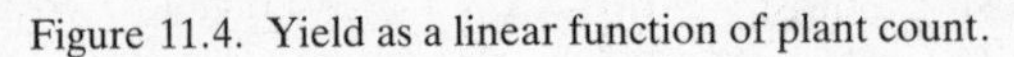

$$Y = 3.24 + 0.028(X) \qquad r^2 = 0.61$$

The function is plotted as a line on the scatter diagram, Figure 11.4. The interpretation of this analysis is that 61% of the variation in crop yield is explained by plant density. The result does not demonstrate that increasing plant density *causes* higher yields. The coefficient β indicates the slope of the line and can be interpreted as follows. An increase of 100 plants per 100 square metres would give rise to a yield response of an additional 2.8 kilograms.

Figure 11.4. Yield as a linear function of plant count.

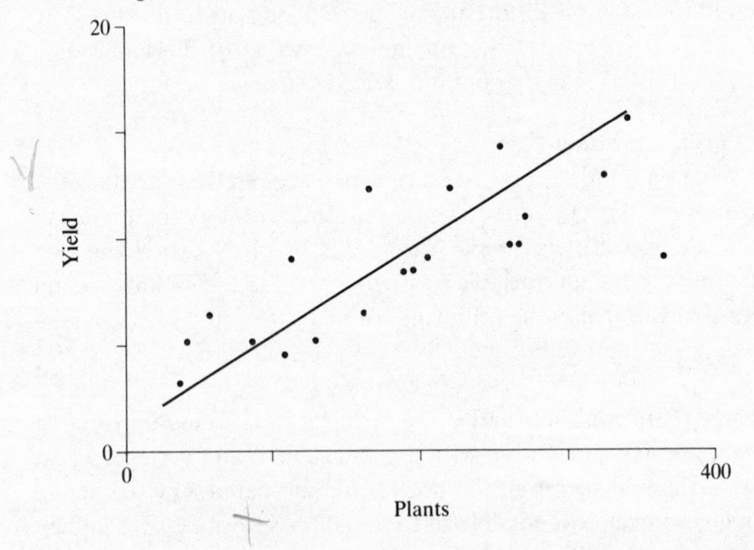

The coefficient α indicates the basal yield level when density of plants is zero. It is clearly irrational that a yield of 3.24 kg could be obtained from zero plants. Interpretation of the regression results is only meaningful within the range of the data used for estimation. Beyond that range the relationship is unknown and could be non-linear. This does not invalidate the model within the specified parameters, but care must be taken over attempts to predict relationships outside that range. The intercept α lies outside the range of data and has no analytical value.

By substituting different values for plant density (X) in the regression equation, predicted yield values (Y) can be obtained. Standard errors can be calculated for both coefficients of regression, and tests of significance made about the values calculated compared with prespecified values.

Regression analysis involves interpreting the data in the form of a model. The functional form of that model will in most cases influence both the finding and the interpretive scope of the analysis. Thus, under the simple linear model illustrated above, there is no limit or maximum level of plant density beyond which yield no longer increases, nor does the rate of yield response (estimated by the β coefficient) vary with the plant density. Theoretical considerations might suggest that more dependent variables should be used, to take account of other factors affecting yield. Equally, if a curvilinear relationship is suspected a different functional form may be necessary. These considerations are outside the scope of this book.

Measuring changes over time

Analysis of change over time is an important objective in many surveys. Sometimes the requirement is for data from just one or two time periods, and a t test of comparison is suitable. But often, especially in the context of project evaluation, the surveyor has as an objective the aim of demonstrating changes in production or income over a number of years, in response to project activities.

Although it is a frequent survey objective there are real problems concerning the length of the data series that is necessary from annual survey data. The objective is to estimate a trend line and test that the slope is significantly greater than zero or a specified value. In order to illustrate the practical implications the process of estimation is considered here.

The functional form of the trend line is a simple linear equation

$$y_i = \alpha + \beta t_i + \varepsilon \tag{11.9}$$

where $\varepsilon \sim N(0,\sigma^2)$, and t_i is the ith time period.

$$\text{Var}\,\beta = \frac{\sigma^2}{\Sigma(t_i - \bar{t})^2} \tag{11.10}$$

where $\bar{t}$ is the mid-point time period.

To measure changes over time the coefficient α represents the base value, and β is the annual increment. The error term ε represents the annual variation. If we are seeking to measure an average annual increase in crop yield of 50 kg/ha, over a base yield of 1000 kg/ha the desired equation would be

$$y_i = 1000 + 50t_i + \varepsilon$$

Our ability to estimate this trend as statistically significant compared with a null hypothesis of zero annual increment depends on four factors

- the annual natural variation in yield
- the size of the annual increment
- the variation in the annual increment
- the length of the time series

These parameters will in practice not be known, but sensible estimates can be used to model the calculation. Consider typical levels of variation

- for annual yield, affected by climate, seed quality and other factors, a coefficient of variation (cv) of 15% of the base yield
- annual increment of 5% of base yield
- for variation in the annual increment, a cv of 40% (equivalent to 2% of base yield)

We require to know the length of the time series necessary to demonstrate a significant trend.

Annual variation is 15% of base yield therefore,

$$\sigma = 0.15y_0$$

$$\text{Variance } \hat{\beta} = (0.02y_o)^2$$

but,

$$\text{Variance } \hat{\beta} = \frac{\sigma^2}{\Sigma(t_i - \bar{t})^2} \qquad \text{from (11.10)}$$

therefore,

$$(0.02y_0)^2 = \frac{(0.15y_0)^2}{\Sigma(t_i - \bar{t})^2}$$

$$\Sigma(t_i - \bar{t})^2 = \frac{0.15^2}{0.02^2}$$

$$= 56$$

We therefore need to use a time series which gives a value of the sum of squared deviations about the mean of t equal to or greater than 56.Table 11.11 illustrates such a calculation.

Table 11.11. *Calculation of deviations of time*

t_i	$\bar{t}$	$(t_i - \bar{t})$	$(t_i - \bar{t})^2$
0		−4.0	16.00
1		−3.0	9.00
2		−2.0	4.00
3		−1.0	1.00
4	4.0	0.0	0.00
5		1.0	1.00
6		2.0	4.00
7		3.0	9.00
8		4.0	16.00
		Sum	60.00

In order to estimate the annual change, β within 2% of base yield a time series of nine points is required. This means data for a base year and eight successive years. In most practical circumstances it is unlikely that data collection could be maintained for such a length of time.

The reader is encouraged to use alternative assumptions about change and variation to estimate the time series required (Casley and Kumar, 1988:118).

11.4 Limitations in analysis

The procedure of analysis, described above, excludes two critical aspects. First, the way in which hypotheses are formulated for testing, and second, the interpretation of test results. Formulation of hypotheses comes from knowledge of the survey subject. That knowledge may be derived from the basic principles of agriculture, economics or sociology, or a detailed understanding of a biological or economic relationship which has been researched and analysed under experimental, or empirical conditions elsewhere. For example, in the study of a draught oxen credit programme described in Chapter 1, one of the objectives was to assess farmer benefits. A review of the literature before the study was planned revealed empirical evidence about changes in crop areas, crop yields and crop output in different farming systems. Hypotheses were therefore set up that there was no difference in these parameters due to oxen ownership. Rejection of these null hypotheses would enable acceptance of alternative hypotheses that a relationship does exist between those parameters and ownership of oxen.

The survey concluded that the evidence supported a strong relationship between both crop area and output, and ownership of oxen but a weak one for crop yield. What do these relationships imply? A survey is not an experiment. Under experimental conditions subjects are chosen to have as many similar characteristics as possible, and are assigned at random to treatment groups. When the treatment is being applied, other factors are held as constant as possible, so that it may be possible to infer that the measured result is caused by the treatment. It is not usually possible to attribute causation from survey data, for a number of reasons.

First, the objectivity of questionnaire surveys must always be held in doubt. A limited number of characteristics can be surveyed by direct, impersonal measurement. But most survey findings emerge from a process of interrogation which is vulnerable to poorly specified questions, careless interviews and distorted replies. The problems of both sampling and non-sampling errors, discussed in Chapter 2, affect both the accuracy and precision of results, as well as the scope for interpretation of findings.

Second, in real populations other influences cannot be held constant. Farm production is influenced by internal factors, such as household labour and skills, and external factors including climate, soils, pests, diseases, and governments. Stratification and other sampling devices can be used to minimise variation from some of these factors, but they cannot be controlled as in an experiment.

Thirdly, in the case of many agricultural processes the interaction of these various influences may not be understood sufficiently well for adequate hypotheses to be studied by questionnaire surveys. A survey designed to measure benefits from a new technology may conclude that gross output has risen. But family and individual earnings will depend on social forces such as land tenure, wage employment, participation by women and ethnic groups. The survey might conclude that production has increased, but is incapable of attributing benefits from that increase without knowing more about the distribution in society.

In the case of the oxen survey a fairly direct causation was argued between crop area and oxen ownership, because ownership of oxen permits a farmer to cultivate his or her own land as early as conditions permit. Under rainfed agriculture with a limited planting period early cultivation will permit a farmer to maximise land area cultivated before planting. Farmers who have to hire oxen can only cultivate when the oxen are available.

For the crop yields it was argued that oxen ownership as such was not a cause of increasing yields. Ownership was thought to enable early and thorough cultivation, leading to a fine seedbed and timely planting. Any yield benefit could be due to these two factors, rather than ownership, because a farmer who had to hire oxen but was able to do so in good time for cultivation could enjoy the same benefits. But equally, the yields recorded could just be due to soils, climate and crop management and not to oxen effects at all. These other factors were not recorded, and were certainly not held constant. The best that could be concluded from the weak yield relationship would be to formulate other hypotheses concerning the actual causation, which could be studied by experiment, or other means.

A final note of caution is for the analyst to beware of over-reliance on test statistics. A statistical test should be used as just one small component of the argument being presented. Conclusions should be drawn from the existence of supporting facts, trends in the data and independent observations, never relying too heavily on isolated statistics (Casley and Kumar, 1988:145).

Summary

Estimates – *not exact* – margin of error – confidence level

Test for differences – reject null hypothesis
possible errors: Type I – commission
Type II – omission

Differences between –
2 estimates – t or z test
multiple classes
discrete data – contingency table
– chi-square statistic
– analysis of variance
continuous data – group by classes (choice of class problem)
– analysis of variance

Relationships between variables
– correlation coefficient
– regression analysis

Change over time – factors – periodic variation
– periodic increment
– variation in periodic increment
– length of series

12

Presenting the results

12.1 Reporting

Content

The purpose of writing a report is to convey information to the reader. This fundamental objective is sometimes forgotten in the enthusiasm to set down as many facts, statistics and deductions as possible. A good report is logically structured, directs the reader to those findings which convey information and draws firm conclusions. If action is called for, the conclusions should be followed by recommendations.

The title should convey a sense of the subject matter and the reason why the report has been written. But the information content of a report has two distinct determinants: those facts which the writer, as analyst of the survey, thinks is important to convey; and what the reader, who may have commissioned the survey, expects to read. Inevitably, these may differ from each other. A basic distinction can be drawn between two different types of survey: a survey with a narrow focus designed to investigate a specific problem or topic; and a more general exercise in data collection for planning or as a benchmark or baseline study prior to measuring change. The reader of a general agro-economic survey would legitimately expect the report to cover fundamental issues including land, labour, capital, production systems, marketing and incomes. The contents of an investigative or evaluation study would be decided by the writer but the reader should be given a clear introduction to the scope and coverage of the study explaining the choice of material to be presented.

Who is it for?

An important early consideration when planning a report is who is going to read it. The more precisely the readership can be defined the easier it will be to direct the content and level of detail to the reader's needs. Reports on public-sector agro-economic surveys generally have four types of reader:

- politicians and senior bureaucrats
- government officers and others with a working involvement
- researchers
- the public

Each group has a different need for information. At the highest level the concern will not be for the detailed findings and arguments, but rather that the study has been done, and that it supports or refutes other proposals or recommendations. Details of findings will be required by those who are professionally involved, who may have to approve the results of the study, and this group will be the most demanding in terms of coverage and thoroughness. By contrast, research workers may share that interest in factual detail, but also request more information about the statistical design, estimation procedures, collection methodology and analysis. Members of the public may not need direct access to the report, but in democratic societies the press may show an interest in the findings and recommendations. It can help this process if a non-technical summary is made available.

To what extent can a single report meet these varying needs? In an ideal situation a different report would be prepared for each group of readers. In practice, this is rarely possible, so the writer must learn to organise the material to satisfy these needs from the one report. Techniques at our disposal include the use of a summary at the start of the document, and appendices for more detailed material. But the main document still has to be presented to a wide range of readers. The only guide for a writer unsure at what level to pitch the content is to write for the most important reader. For any survey report it is usually possible to identify one reader whose response to the document has paramount importance: it may be a project manager, permanent secretary, committee chairperson or funding agency representative. Whoever it is, write the report for their needs and accept the minor criticisms from readers whose needs are not so fully met.

Circulation

Circulation of a final report will usually be decided by the organisation which commissioned the study, and will be based on established practice. The writer may have little influence over the distribution list. Wherever possible, it is good practice for a copy to be sent to all organisations or individuals which gave assistance, even if the final document is of little apparent relevance to them. It is an act of simple courtesy and a suitable way to thank people for their cooperation.

Distribution of final reports is self-evident. Distribution of interim or draft reports is more of a dilemma. There is undoubtedly much to be gained by circulating working drafts for comment on style and contents. Report writers should always be mindful of the need for timely reporting, and a working draft is an indication that progress is being made. However, there is always a danger that statistics will be recalculated after checking for errors, or under different analytical assumptions, with significant changes to the order of magnitude. If report readers act on preliminary data, only to find revisions which change the sense of the conclusions, there will be frustration at the writer and the survey organisation. As a guide:

- preliminary data should only be circulated amongst people who are closely involved in the study
- a better policy is to await final calculations and circulate for comment on the narrative and conclusions

But equally, it is important not to wait until a final report, especially if the writer has been commissioned to do the study. Draft texts allow the writer to respond to suggestions by the sponsor and so the sponsor becomes identified with the contents, rather than being a passive recipient. This is especially important to avoid petty criticism from recipients who feel they have not been consulted about the contents.

Summary

Report	– to convey information
content	– writer's interpretation
	– reader's expectation
readership	– write for the main reader
circulation	– when data are finalised for comment on text

12.2 Report structure and contents

We wrote above that the purpose of writing a report is to convey information to the reader. There are distinct features about a report which make information easy to learn and to remember. Even if you are working in a language which is not your native tongue, and without sophisticated word processing and printing facilities, attention to a number of basic principles will improve readability and comprehension of the report. These features are described under five headings, the five Ss.

- style
- structure
- signposts
- stress
- summaries

Style

Writing style is perhaps the most elusive of the list, and to inexperienced writers the most worrying. The style of a document creates a relationship between the writer and the reader, and acts as a conductor or a barrier to interpretation. But what is meant by style and what constitutes good style? The references in the bibliography include an analysis of writing skills which stress the need for simplicity of construction and clarity of expression (Cooper, 1975).

A basic feature of sentence construction is word order. There is a clear difference of emphasis in these two sentences.

> Of the specific agronomic recommendations accompanying the improved sorghum varieties, that on sole cropping is shown by Tables 2.3(A)–(C) to have had very little impact.
>
> Sole cropping of the improved sorghum varieties, as a specific agronomic recommendation, is shown by Tables 2.3(A)–(C) to have had very little impact.

Although the usual way to construct a sentence is subject first, followed by verb and then object or complement, greater emphasis may be achieved by inverting the order. What is the subject of the sentence above? Clarity is aided by keeping sentences short and simple, avoiding complex constructions. The quotation above tells us that sole cropping of sorghum was a recommendation *and* that it had little impact on the farmers' practices. Separate sentences might have been better.

> Sole cropping was one of the specific recommendations which accompanied the new sorghum varieties. The survey data in Tables 2.3(A)–(C) show that this had little impact.

The readability of text is improved if sentence length is varied, mixing short, simple statements with longer constructions. Where possible, avoid long words, foreign expressions, scientific terminology and jargon. What is meant by 'had little impact'? If technical or specialised expressions are required for precision, explain them when they are first introduced, and make use of a common name or equivalent thereafter. Try to write in the active sense and avoid the passive and impersonal style of legal documents. The active voice gives a different emphasis and requires fewer words. It also sounds less pompous. A simpler version of the quotation might be:

> Sole cropping was recommended for the new sorghum, but few farmers took this advice (Table 2.3).

Style is personal, and cannot be prescribed by a set of rules. Widely differing styles, such as those of Ernest Hemingway or James Clavell can be equally appealing. Gripping prose is not called for in a survey report, but an attractive style is a help for the reader. More important though, is clarity. *The Economist* newspaper prides itself on good writing. In its style book it argues that clear thinking is the key to clear writing. '. . . think what you want to say, then say it as simply as possible.' This will be easier if the report is set out in a logical structure (*The Economist*, 1986).

Structure

Report structure will vary according to the type of document being written and the scope of the subject matter. A one-page memo reporting a short field visit will be set out in a different format from a survey analysis of a year's data. Some features are common to all reports. A basic principle is to have a beginning, a middle and an end. For the survey this can be interpreted as 1. a statement of purpose in the survey 2. presentation of the findings and interpretation, and 3. a conclusion.

With a large survey, the presentation of findings may take several sections or chapters, and the sequence of material here is important. Think of the readers, especially if they are likely not to be familiar with the report subject. Choose a structure which follows a natural progression, such as a chronological sequence, or the steps in a process. If there is no natural order, devote space to explaining the order of presentation first. If the subject is complex a good starting point might be the hypotheses or objectives on which the survey was designed, even if the subsequent findings contradict them.

The sequence of operations starts with the material to be presented. This will generally require results from the survey, other published data and evidence from similar studies and reports. Because most reports need information that is not collected as part of the study it is important to consider report preparation at an early stage in the survey process. If this stage is omitted the final report may be delayed while other material is being located.

Once the material is assembled the final structure can be agreed. Presentation of findings usually consists of a statement of results, often a table or diagram, followed by a discussion. One of the challenges facing a writer is to minimise the clutter of data while still presenting enough information for the argument. Presenting data and graphs is covered in later sections.

Report structure may be at the discretion of the writer, or there may be a convention to be followed. Example 12.1 presents a general-purpose layout.

Physical layout is also an aid to interpretation. Start new sections on new pages. Narrower text (like newspaper columns) is easier to read so keep lines short (10–11 cm) with wide margins. Hanging paragraphs, where the paragraph number or first words start offset in the left margin helps readers to find sections quickly. For reports with long sections of text use double spacing, but vary the spacing for summaries and quotations.

EXAMPLE 12.1

Report structure and contents

Title page

Summary

This should be a review of the contents, including conclusion and recommendations, in the same order as the report. It should not contain any material or deduction not mentioned in the main body. In a long report it is helpful to reference the location of topics in the main body.

Introduction

To state what the report is about, background information, who it is written for and how the subject is dealt with.

Methodology

Give a general overview in the main text, indicating the material which is unique to the study, and the use made of existing or secondary data. For a major study, or where complex techniques have been used, fuller details should be given, but they are probably better placed in an appendix. Coverage:

- The reference and recall periods for material.
- Survey design and execution.
- Full details of samples, e.g. a two-stage sample with PPS selection of cooperatives at the first stage and a LSS of dairy farmers at the second stage. Note and explain the use of stratification.
- Give details of sample units that were dropped or added after the initial sample. Give reasons and dates.
- List or make reference to the sample frames used. This is important to assess the need for weighting.
- Include in an appendix, or describe, the questionnaires and other survey documents.
- Use of special or unusual analytical techniques.

Main body of text

Adopt a logical approach, as described above. Relate findings to practical aspects of development, and draw conclusions from the evidence at each stage of the text.

Conclusion and recommendations

This can be presented as one or two sections, depending on the purpose of the report. The conclusion should not contain any ideas not previously raised in the report. It should state in unqualified terms what the findings are. The findings should then be related to other evidence or hypotheses as described in the main body, including a comparison with other, similar studies. Where appropriate, a variety of options for action arising from the study should be explained.

References

Appendices

Signposts

The reader will find it easier to follow the presentation of material and arguments if there are clear directions around the report. A basic requirement is a table of contents. Even better is an index, which is now a feature of some word processing programmes. If an index is impractical the table of contents should be sufficiently detailed to guide the reader to individual topics.

The text is easier to follow if it is divided by headings and sub-headings, with a consistent typographical style of upper and lower case letters, underlining, and boldface type. Different typefaces and letter sizes are also possible with many electric typewriters and computer printers. The heading scheme should be based on the divisions used in the report. It is better to restrict subordinate headings to two or three levels. If more levels are needed the main headings may be too broad and should be expanded.

Numbering of sections *or* paragraphs will help the reader, and is an aid to cross-referencing. The decimal system is simple and one of the most popular. Alternative schemes use Roman numerals or combinations of letters and numbers.

Headings are numbered through the report.

Chapter or section	1
First level subsection	1.1
Second level subsection	1.1.1

If paragraph numbers are used it is a good idea to number within main sections. The numbers stay smaller and changes are easier to make if the text is edited. Again, the decimal system is popular.

Chapter or section	2
First paragraph	2.01
Second paragraph	2.02

Paragraph numbers are particularly helpful for cross-referencing and easier than page numbers for the writer to use before the report is finally printed.

Tables, diagrams and figures can be numbered within the main section (if there are many) or consecutively through the report.

Stress

At various points in the narrative the writer will want to emphasise a statement or conclusion, to make it stand out. Simple devices can be used here: varying typeface; boldface; underlining; line spacing; use of wider margins (narrower text); and boxes. Except for changes in typeface these devices are equally accessible to a writer whose report is prepared on a manual typewriter and paper stencils, as to a writer using a word processor and laser printer. Some writers may question the need for such devices. When you are intimately involved with the subject of the report it is easy to forget that for many readers the contents are unknown, perhaps complex. Simple devices to stress words, phrases and illustrations help vary the pace of the report and keep the reader's attention. Some good examples of this can be found in magazines and newspapers, such as *The Economist*.

Readers in a hurry, who tend to skim pages, are often drawn to pictures or diagrams. The captions to these are more likely to be read than the accompanying text, so they should be used to convey important facts or conclusions.

Summaries

Make regular summaries during the report: after each section of facts and new interpretation; at the end of each main section; and in a section to describe the whole report. Summaries reinforce the argument being made and simplify key points in a more memorable way. With the use of intermediate summarising paragraphs it is easier for the reader to retain different strands of the evidence and to follow the arguments in a conclusion.

Summaries do not have to be in prose. Lists are a simplifying presentation, and are often used as an introductory device to a set of topics. Some readers like the flow diagram device used in this book. This idea, which comes from a book by Tony Buzan, called *Use Your Head* is intended to show the logical framework relating concepts together spatially, and to avoid the linear style of prose and numbered lists.

Summaries are best made after presenting facts and arguments, but for reports which have to satisfy different readers a summary section at the beginning of the report and to introduce each main section, will help a casual reader to decide whether to look for more detail or skip to a later section.

Summary

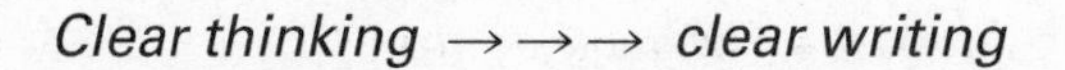

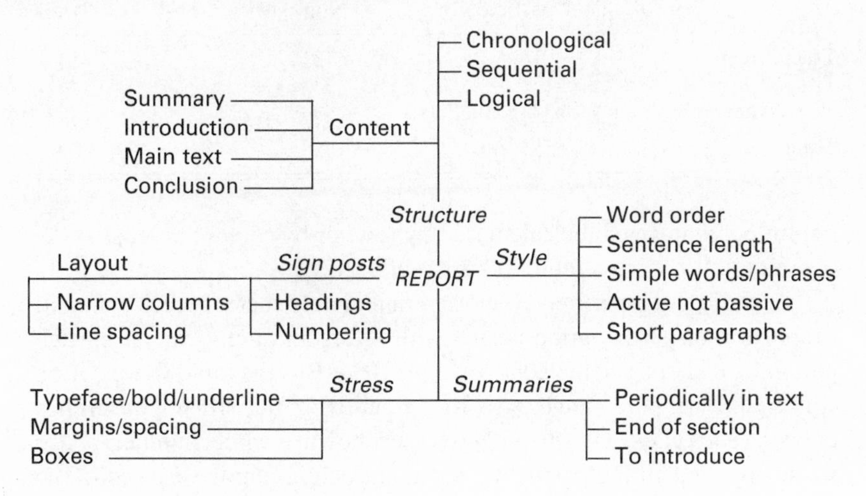

12.3 Presenting data

Two-way tables to classify variables

Throughout Chapters 10 and 11, use has been made of tables to illustrate exploratory data analysis and the techniques available to test for relationships between variables. The two-way table or cross-tabulation is one of the most popular and adaptable methods of presenting data, and is widely used in reports. In this chapter we look at the construction of tables and ways of simplifying and improving their presentation. A thorough and readable account of data presentation can be found in Chapman and Mahon (1986).

A two-way classification was used in Chapter 11 for the analysis of households by tractor ownership and participation in extension. The table reported actual counts and is repeated here (Table 12.1).

We can see that a minority of farmers participate in each category, but it is not clear if the proportions are similar or very different. All cross-tabulations of this type are better presented in percentages calculated from whichever variable is of greater importance. In this case, our concern is with tractor ownership and how it affects extension participation so we will calculate percentages for each class of tractor ownership, by extension category. Computer statistics software will normally

Table 12.1. *Number of households by tractor ownership and extension*

	Extension participator	Non-participator	Total
Tractor owner	15	131	146
Tractor hirer	152	335	487
Non-owner or hirer	384	990	1374
Total	551	1456	2007

permit percentages to be calculated by row, or by column or against the overall total, for cross-tabulations (Table 12.2).

Calculation of percentages shows immediately that whereas a similar proportion of participation occurs with hirers and non-owners, participation by owners is some 20% lower at 10%. But the table is now more cluttered. The percentages are too precise for the size of difference between each class. They would be better shown as whole numbers. Also we do not need both the percentage and the actual count in each cell. The percentages give a better comparison between the classes but we do need to be able to check the actual number of observations to know if any of the categories has a low representation. For example, if our survey recorded only 20 or 30 tractor owners out of 2007 farm households, our readers may be less confident of our conclusions. An efficient solution is to quote percentages in the body of the table, and indicate number of observations at the row or column totals, as in Table 12.3. In this way the reader has a clear presentation, but can check on sample sizes if necessary, by applying the percentage to the total.

The quoting of percentages without sample size is used to good effect in product advertising. A current advertisement for cat food in the UK

Table 12.2. *Households by tractor ownership and extension (row percentages shown in brackets)*

	Extension participator	Non-participator	Total
Tractor owner	15	131	146
	(10.27)	(89.73)	(100)
Tractor hirer	152	335	487
	(31.21)	(68.79)	(100)
Non-owner or hirer	384	990	1374
	(27.95)	(72.05)	(100)
Total	551	1456	2007
	(27.45)	(72.55)	(100)

Table 12.3. *Percentage of households by tractor ownership and extension*

	Extension participator	Non-participator	%	Number of households
Tractor owner	10	90	100	146
Tractor hirer	31	69	100	487
Non-owner or hirer	28	72	100	1374
Total	69	231	100	2007

claims that seven out of ten cat owners who expressed a preference chose the advertised product. But we are not told how many owners expressed a preference. If we were to find that most of the sample had no preference at all it could undermine our confidence in the true popularity of that brand. Happily, neither of the authors buys cat food.

Row percentage, column percentage, or overall percentage?

In the example above, percentages were calculated by row. Each cell is expressed as a percentage of the row total. Percentages could have been calculated by columns, or by the grand total of the table. Computer software will often include a choice of any one or all three calculations. Interpretation of the data will change according to the percentage calculated so the choice of percentage must be made with care. If the sample is stratified, or divided into domains, it is usual to calculate percentages within strata. In the example the dividing characteristic was tractor ownership or use, and extension was an independent classification, so percentages were calculated for each class of tractor ownership.

Frequency lists

Our next example concerns a frequency distribution in the form of a list. Table 12.4 presents a classification of farms in the European Community with the percentage in each class with access to an Outside Gainful Activity (OGA). The purpose of showing the table was to identify those types of farming with the highest relative frequency of OGA. The way the data are presented the reader has to scan a list of 17 types in order to check the types of farming with OGA frequencies higher than the share of holdings in that category.

Simply by ordering the table on the basis of OGA proportions, and dividing the list of types of farming, the presentation is improved (Table 12.5). It is readily seen that eight types of farm have a higher proportion

Table 12.4. *Classification of farms*

Type of farming	Farm holders with OGAs (%)	All farm holders (%)
Cereals	11.30	9.20
Field crops, other	12.70	12.60
Horticulture	1.70	2.10
Vineyards	11.60	9.70
Fruit/permanent crops, other	20.50	17.80
Cattle, dairying	6.80	11.20
Cattle, rearing/fattening	4.00	4.00
Cattle, mixed	1.50	2.30
Grazing livestock, other	6.50	6.20
Pigs	0.80	0.70
Pigs and poultry, other	0.50	0.50
Horticulture and permanent crops	0.20	0.20
Fixed cropping, other	10.60	10.40
Partially dominant grazing livestock	2.50	3.10
Fixed livestock, other	0.60	0.80
Field crops and grazing livestock	5.40	6.30
Crops-livestock, other	2.80	2.60
All types	100.00	100.00

Source: Eurostat, Reported in JAE 38, 173

Table 12.5. *Ordered classification of farms*

Type of farming	Farm holders with OGAs (%)	All farm holders (%)
Fruit/permanent crops, other	20.50	17.80
Field crops, other	12.70	12.60
Vineyards	11.60	9.70
Cereals	11.30	9.20
Fixed cropping, other	10.60	10.40
Grazing livestock, other	6.50	6.20
Crops-livestock, other	2.80	2.60
Pigs	0.80	0.70
Cattle, dairying	6.80	11.20
Field crops and grazing livestock	5.40	6.30
Cattle, rearing/fattening	4.00	4.00
Partially dominant grazing livestock	2.50	3.10
Horticulture	1.70	2.10
Cattle, mixed	1.50	2.30
Fixed livestock, other	0.60	0.80
Pigs and poultry, other	0.50	0.50
Horticulture and permanent crops	0.20	0.20
All types	100.00	100.00

of OGAs than their proportion amongst all holders. The interpretation is that OGAs are more frequently found amongst those farm types with a lower labour requirement. The data have not been changed but this conclusion follows more readily from the second table than from the first.

Four principles can be derived from these examples:

1. Try to present distributions in percentages to enable comparison between categories from different subpopulations.
2. Always show the sample size for the category and overall, when quoting percentages in a table.
3. Round percentages to integers unless the analysis requires comparison of small differences.
4. When presenting lists and response frequencies, order or group the list according to the points you wish to explain in your interpretation.

Descriptive statistics

Tables are used in a report to simplify the interpretation of data. This is best illustrated, using the principles set out above, by starting with a listing of data collected from a number of households (Table 12.6). The purpose of the survey was to estimate average area of each crop, average output and the proportion using fertiliser.

The data show the area of major crops grown by a sample of ten households. All the data are present, but interpretation is difficult because of the way they are grouped, and the amount of detail provided. The first step to a better presentation is to group or order the data in a better way. If the analysis is being done by hand this next step would make use of a tally sheet, described in Chapter 9. We will rearrange the data by crop type (Table 12.7).

This simple rearrangement brings an immediate improvement and is similar to an ordered tabulation sheet, but the reader still has to check every observation to understand the results. The next step after grouping the data is to prepare sample counts or averages for each category (Table 12.8).

This third table shows the count of observations for each crop; crops are presented in descending order of average crop area. From a raw data listing we have now progressed to see that on average cotton occupies the largest area at 0.60 ha, compared with sorghum and maize, both of which are about 0.40 ha, followed by groundnuts at 0.13 ha.

This table now illustrates a common problem facing analysts, and one which is often not dealt with very well. The averages calculated here are

Table 12.6. *Household crop production data listing*

Household number	Crop	Area (ha)	Output (kg)	Fertiliser use Y/N
1	maize	0.76	1250	Y
1	cotton	0.25	310	
1	groundnut	0.15	150	
1	sorghum	1.08	1400	
2	maize	0.25	450	
2	groundnut	0.40	250	
3	cotton	1.75	2150	Y
4	maize	0.60	1150	Y
4	cotton	0.60	870	Y
4	groundnut	0.20	175	
5	maize	0.75	1440	
5	sorghum	0.75	1200	
6	cotton	1.10	1150	
6	sorghum	0.55	1050	
7	maize	0.18	200	
7	groundnut	0.25	210	
7	sorghum	1.10	1970	
8	cotton	1.65	2500	
9	maize	0.55	750	Y
9	groundnut	0.25	100	
9	sorghum	0.55	250	
10	maize	0.90	1760	Y
10	cotton	0.65	1000	Y

for the whole sample. But we can see that only six households out of our sample of ten grew cotton, five grew sorghum and groundnut, and seven grew maize. What meaning should be attached to an average of all ten which includes non-growers? Similarly, although three farmers used fertiliser on cotton and four on maize, giving sample proportions of 30% and 40% respectively, the proportion of *growers* who used fertiliser was in fact higher, at 3/6 = 50% for cotton, and 4/7 = 57% for maize.

The analyst faces two decisions here:

- which data should be reported for the whole sample and which for subpopulations only
- how best to convey neatly both sets of statistics.

A descriptive survey will usually require all results to be presented for the survey population, thus averages and counts would be based on the whole sample. But studies designed to investigate farmer behaviour, such as response to extension programmes or the performance of new technology, such as seeds and fertiliser, will be more concerned with those farmers participating in the programme, or growing the crop. Results for

Table 12.7. *Household crop production – grouped data*

Household	Crop	Area (ha)	Output (kg)	Fertiliser use Y/N
1	cotton	0.25	310	
3	cotton	1.75	2150	Y
4	cotton	0.60	870	Y
6	cotton	1.10	1150	
8	cotton	1.65	2500	
10	cotton	0.65	1000	Y
1	groundnut	0.15	150	
2	groundnut	0.40	250	
4	groundnut	0.20	175	
7	groundnut	0.25	210	
9	groundnut	0.25	100	
1	maize	0.76	1250	Y
2	maize	0.25	450	
4	maize	0.60	1150	Y
5	maize	0.75	1440	
7	maize	0.18	200	
9	maize	0.55	750	Y
10	maize	0.90	1760	Y
1	sorghum	1.08	1400	
5	sorghum	0.75	1200	
6	sorghum	0.55	1050	
7	sorghum	1.10	1970	
9	sorghum	0.55	250	

Table 12.8. *Household crop production – average data*

Household count	Crop	Sample mean Area (ha)	Sample mean Output (kg)	Fertiliser use (count of Y)
6	cotton	0.60	798	3
5	sorghum	0.41	587	0
7	maize	0.40	700	4
5	groundnut	0.13	89	0

Table 12.9. *Crop areas and output for growers only*

Household growers %	Crop	Sample mean Area (ha)	Sample mean Output (kg)	Fertiliser use (% by growers)
60	cotton	1.00	1330	50
50	sorghum	0.81	1174	0
70	maize	0.57	1000	57
50	groundnut	0.25	177	0

this second type of analysis would be based on the subpopulation with that characteristic. Continuing with the same example, Table 12.9 illustrates how different the average areas and output are when they are calculated for growers only, rather than across the ten households.

The interpretation of the two tables would differ like this.

> From the survey, the average area of cotton grown in the population was found to be 0.60 ha. However, just 60% of households grow cotton and amongst those households the average area was 1.00 ha. Half the cotton growers applied fertiliser to their crop

The average for the population would be used to calculate total cropped area for the study. But if farm budgets are to be prepared for different farm types the average of growers would be the correct figure to use.

Mean X *or Mean* N

The terminology used so far is clumsy: the mean for the whole sample and the mean for units of study which possess the characteristic (growers, in the example). A simplifying terminology is to refer to Mean X or Mean N. Here we make use of some simple notation: N is a common symbol for sample size, so Mean N is the mean over the whole sample. X refers to those observations which possess the characteristic, so Mean X would be the mean for growers in the table.

Distribution and central tendency

When we considered exploratory analysis in Chapter 10 it was stressed that the distribution and central tendency of data should be examined as a preliminary part of analysis. Yet the examples which followed in Chapter 10 and in previous sections in this chapter have been concerned primarily with tables of counts and averages, with little mention of dispersion. A problem facing all reporters is the balance between a comprehensive display of statistics and clarity of presentation. The cross-tabulations we have used as examples indicate the simple relationships between variables but tell the reader nothing about the variables themselves. For each variable the analyst needs to show as a minimum:

- the mean (N and/or X as necessary) or other central statistic
- the dispersion about the mean

– the existence of any serious skewness

The problem is how to achieve this and keep the presentation attractive.

As a general principle the main body of the report should contain only those statistics which are essential to the analysis or argument being presented. All non-essential descriptive statistics should be presented in an appendix. The exceptions would be where frequency distributions or other features have such an important bearing on interpretation that presentation in the main report is unavoidable. Appendix tables should present sufficient detail for interested readers to review the statistics and make their own judgement about the values collected.

The layout of data in an appendix must inevitably depend on the purpose and topics of the data, but a minimum set of statistics would include the mean, standard deviation, standard error, maximum and minimum, together with sample size, as in Table 12.10.

It will not always be convenient to confine statistical details to an appendix. Many reports are intended for readers who would expect full coverage of descriptive statistics, yet the writer still must try to avoid long, detailed tables. A simple solution is to adopt a symbolic code in a similar fashion to the method used for showing the level of significance of variables. The purpose of reporting the standard deviation is to indicate the level of variability, as described in Chapter 10. This can be presented directly as the standard deviation, or as the standard deviation expressed as a percentage of the mean. But for some readers a more useful presentation is either the standard error, or the standard error as a percentage of the mean. If the standard error is known the confidence interval can be quickly computed from the z or t value. At 95% probability the confidence interval is twice the standard error (rounding $z = 1.96$).

Table 12.10. *Appendix data tables – bundle unit weight for coffee (tonne)*

	X	Min	Max	Mean X	Mean N	SD N	SD %	SE N	SE %
Area 1	35	0.04	0.38	0.20	0.20	0.09	45	0.015	8
Area 2	27	0.12	0.32	0.27	0.21	0.06	29	0.010	5
Area 3	32	0.13	0.52	0.27	0.25	0.08	32	0.014	5
Area 4	21	0.05	0.11	0.15	0.09	0.03	33	0.005	6
etc.									

Survey sample (N) = 35

The range within which either the standard deviation or standard error lies can be indicated by a symbol. The list below shows a sensible range for survey data. A variable with a very high standard deviation would not have an asterisk next to it. The code follows the familiar pattern of 'the more blobs the better'.

*	standard deviation >25% and <51% of mean
**	standard deviation >10% and <26% of mean
***	standard deviation <11% of mean

A similar system can be devised for the standard error.

~	standard error >10% and <21% of mean
~~	standard error >5% and <11% of mean
~~	standard error <6% of mean

The two code systems are illustrated by an example of mean hours worked per hectare on the cultivation of four crop enterprises (Tables 12.11–12.13).

Both code systems are an improvement over the original table, which contains too many numbers for easy interpretation. The symbols for standard deviation indicate that most of the data are very variable and

Table 12.11. *Average labour input (Mean hours, standard deviation shown in brackets)*

	Sorghum	Sorghum + millet	Cotton	Cotton + millet
Land clearing, preparation and planting	56.17 (27.23)	57.51 (39.05)	61.56 (44.00)	77.69 (25.34)
Post planting	180.49 (76.00)	204.47 (109.61)	164.53 (87.10)	319.57 (222.13)
Harvest	93.37 (68.11)	115.17 (35.54)	156.09 (35.21)	211.69 (46.66)
Post harvest	10.75 (9.87)	15.58 (10.45)	8.36 (7.77)	23.38 (12.57)
Sample size (plots)	194	184	45	27

Table 12.12. *Average labour input (Mean hours, standard deviation indicated by *)*

	Sorghum	Sorghum + millet	Cotton	Cotton + millet
Land clearing, preparation and planting	56*	58	62	78*
Post planting	180*	204	165	320
Harvest	93	115*	156**	212**
Post harvest	11	16	8	23
Sample size (plots)	194	184	45	27

have a standard deviation greater than 50% of the mean. To identify the precision of the data, knowing the standard deviation, the sample size is needed as well, so it is included at the foot of the table. The third table (Table 12.13) indicates the level of precision and includes the effect of sample size, which is now omitted from the table. This example has assumed a simple random single stage sample. With a more complicated sample design the correct formulae for standard error should be used and it will probably be more convenient to express the variation in terms of the standard error. Both coded tables are improved by the data being rounded to integer values. Unless small differences between values are important tables are better presented with numbers rounded to the nearest whole number.
category contains a response of 2% greater than K1000. Unless the actual values are known it will not be possible to set an upper limit for this class on the graph. In order not to distort the histogram it may be necessary to show a break in the income axis and indicate higher values beyond the break (see Figure 12.2).

Care must be taken when plotting histograms using computer software. Many programmes which appear to plot histograms actually draw bar charts which can be adapted to histograms. The graphs are drawn to a linear scale with the assumption that class intervals are equal. With bar chart software the example above would need to be specified entirely in K50 classes to be graphed correctly as a histogram with an area scale.

Table 12.13. *Average labour input (Mean hours, standard error indicated by ˜)*

	Sorghum	Sorghum + millet	Cotton	Cotton + millet
Land clearing, preparation and planting	56˜˜˜	58˜˜˜	62˜	78˜˜
Post planting	180˜˜˜	204˜˜˜	165˜˜	320˜
Harvest	93˜˜˜	115˜˜˜	156˜˜˜	212˜˜˜
Post harvest	11˜˜	16˜˜˜	8˜	23˜˜

12.4 Graphs

Bar chart

The bar chart is a simple and effective way of improving the presentation of response frequencies for ordinal or classificatory variables. A bar, or column, is used to indicate the response or value, either by a count or as a percentage. For greater visual impact the bar can be replaced by symbols derived from the subject of the table. These diagrams are used widely in newspapers and magazines where clarity of expression and rapid interpretation are essential to hold the reader's interest. Variation in shade and the use of colour can be used to highlight features. For summaries of key findings a large scale, with the response category written along the bar, can be very effective. An example using labour data is shown in Figure 12.1.

Histogram

The histogram differs from the bar chart in the way the frequency of response is presented. The bar chart uses a linear scale. The longer the bar the greater the value. The histogram is based on an area scale where both length and width are important. The histogram is used to present continuous data which are grouped into classes such as the frequency distributions built upon the data in Table 11.8. When drawing a histogram care must be taken over the class intervals and the treatment of outlying values.

In the example in Table 12.14 the frequency classes are not equal because the number of responses above K(kwacha)250 were low and

Figure 12.1. Labour utilisation on four crop enterprises.

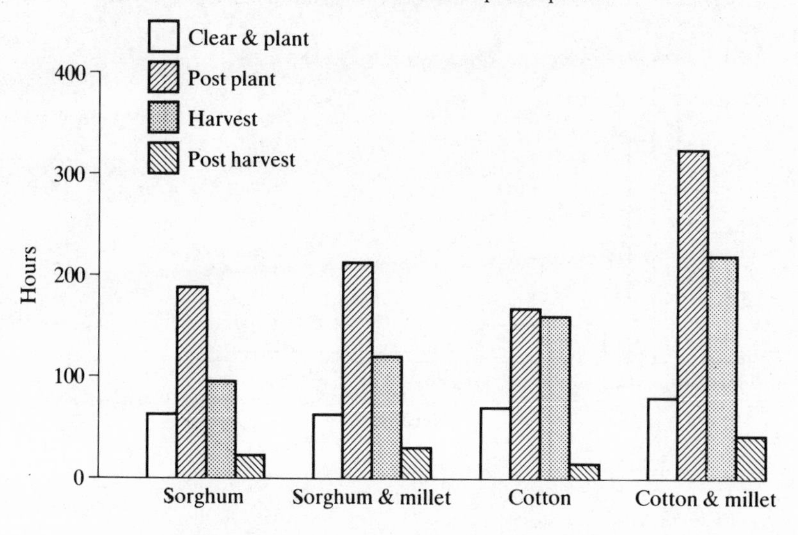

Table 12.14. *Frequency distribution – households by income class*

Income class (kwacha)	Households (%)
0–<50	12
50–<100	19
100–<150	24
150–<200	17
200–<250	6
250–<500	15
500–<1000	5
≥1000	2

scattered over a wide range. When drawing a histogram from the data the width of each column must be proportional to the width of the class interval. Thus the width for the classes K250 to K500 and K500 to K1000 must be five times and ten times respectively the width of the earlier classes, which are K50 wide. The percentage response must also be spread over that wider range. A response of 15% spread over five classes should be graphed as 3% per class. Similarly, the 5% reponse in the range 500 to 1000 is half a per cent for each K50 class. In this way the area of each column is proportional to the frequency of response. The final

Figure 12.2. Histogram of household income distribution.

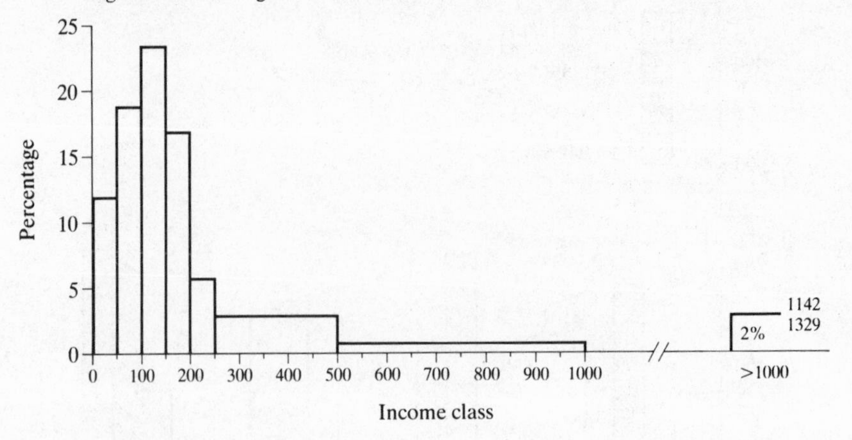

Pie chart

The pie chart offers an alternative style of presentation for frequency of response data. It is most suited to ordinal scales of measurement. Computer software will permit contrasting shading of each segment and often includes the facility to 'explode' or 'extract' a segment in order to draw attention to the category. The example in Figure 12.3 is derived from national budgetary allocations.

Figure 12.3. Papua New Guinea. Budgetary allocations by sector.

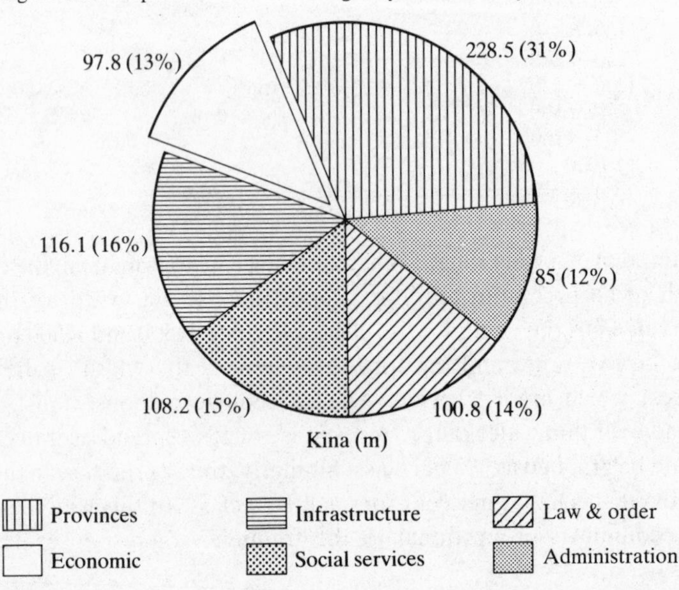

Line graph

The line graph is the most adaptable style of diagram and can be used for a wide variety of data including frequency distributions, response curves and time series plots. The choice of scale for the vertical axis can have a major impact on the look of the data and therefore on the interpretation placed on it by the reader. The two graphs in Figures 12.4 and 12.5 show annual data on different scales. The first graph conveys an impression of great variation, with large changes in values from year to

Figure 12.4. Farm cereal area.

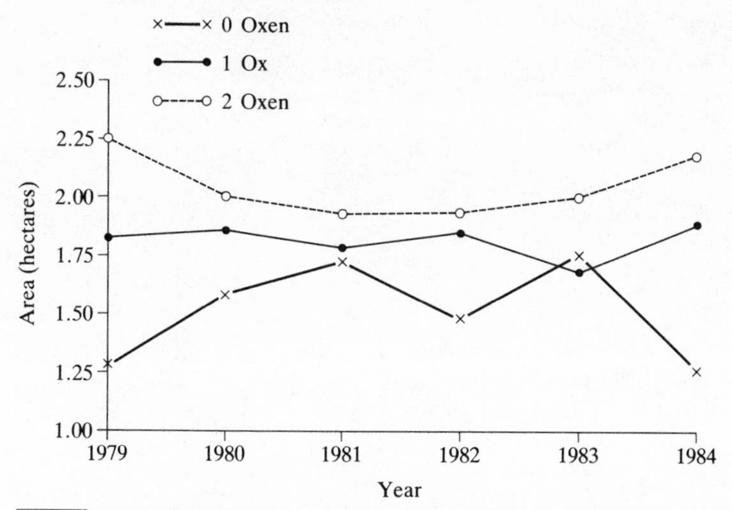

Figure 12.5. Farm cereal area.

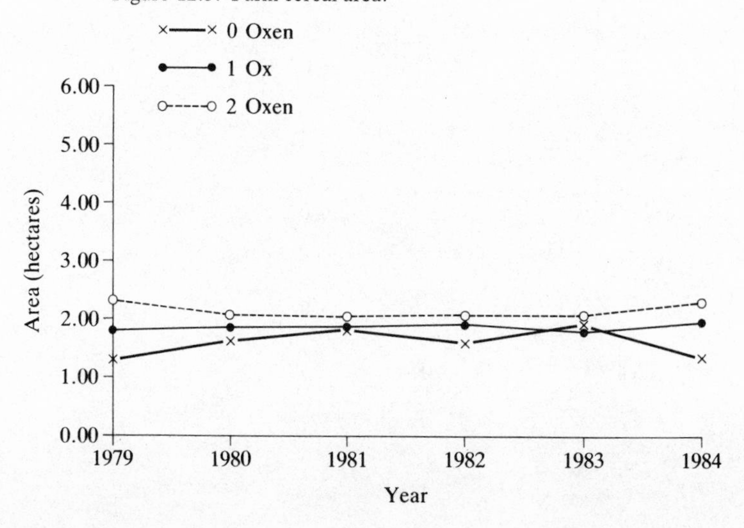

year. The second appears to be a mild fluctuation about a stable trend. Both contain the same data.

High–low chart

The final example, in Figure 12.6, is of a High–Low chart which indicates a maximum, minimum and average value for data in a series. A simple refinement would be to mark one standard deviation either side of the mean. The illustration of price data is particularly suited to this presentation.

Figure 12.6. Wholesale price of wheat.

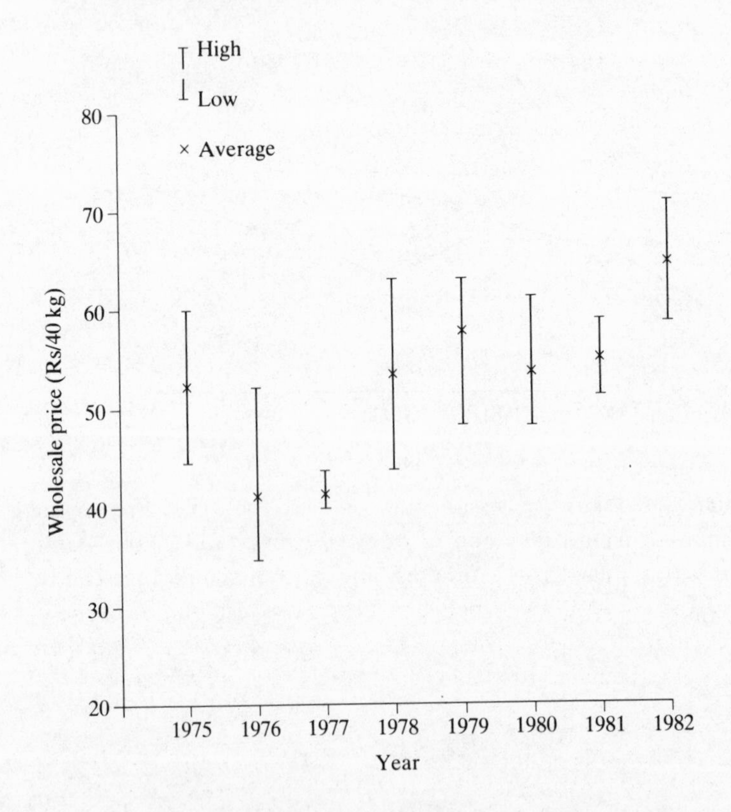

Appendix 1

Estimating parameters

This Appendix should be read in conjunction with Chapters 3 and 10. It contains worked examples of the calculation of population totals, means and ratios from three common sample designs:

- a single-stage simple random sample
- a two-stage PPS/SRS sample
- a two-stage SRS/SRS sample

The material here uses simplified numerical examples to illustrate calculations.

In many surveys the analyst will be content to present results as averages for the sample data, without any attempt to calculate population estimates. But if the purpose of the survey is to quantify information about a specific population – a cooperative, a project area, or a region – population estimates of totals and averages, with their standard errors, will be required. The calculation of standard errors will enable confidence limits to be placed on estimates, and tests of significance can be carried out on the difference between estimates. Chapter 11 reviews methods of analysis which make use of population estimates and their standard errors calculated by the methods shown in this Appendix.

Calculating population estimates from sample estimates

Population estimates are fundamental to any survey analysis. The basic approach can be illustrated independently of statistical design. Totals, means and proportions are calculated from the survey data. They are then raised from an estimate of a small sample to an estimate of the population by multiplying by the inverse of the sampling fraction.

EXAMPLE A1.1

Raising sample total to population total

In a village of 1000 households, a sample of ten households was drawn. The sampling fraction is 10/1000 or 0.01. If rice production was measured for each household in the sample we can calculate total sample household rice production by summing each household's production. To calculate an estimate of total rice production for the whole village the sample figure must be 'raised' to the population. The factor necessary to do this is equal to the number of times by which the sample would have to be multiplied to equal the population. That is 1000/10 or 100. This is the inverse of the sampling fraction. So if the sample rice production equalled 22 tonnes, production for the population (the whole village) would be 22 tonnes multiplied by 100, giving 2200 tonnes.

This principle holds true for all estimation. Complications arise from the sample design. The more complicated the design in terms of number of stages, variation in sampling fraction and sampling with or without replacement, the more complicated will be the algebra for the calculations. This is especially true for estimating variance, and for many designs access to a computer is desirable if variance is to be calculated for many variables.

The following examples start with a single-stage simple random sample and progress to two illustrations of two-stage samples. Of necessity the simplest cases with self-weighting designs are used. If more complex designs are needed the surveyor should seek professional advice from a statistician. Some algebra is unavoidable, but wherever possible a description accompanies the symbols.

Calculation of total, mean, ratio and proportion from a single stage Simple Random Sample

Using the following notation:

Value of variable *Y* for *i*th sample unit	y_i
Number of units in sample	n
Number of units in population	N
Sampling fraction	$f = n/N$
Sample mean of *Y*	$\bar{y}$
Population total of *Y*	Y

Population mean of *Y*	$\bar{Y}$
Population proportion	P
Sample proportion	p
Population ratio	R
Sample ratio	r
Number of units in population possessing a characteristic	A
Number of units in sample possessing a characteristic	a

Notation for a second variable X is similar to that for Y.
Estimates are indicated by a caret over the symbol, thus the estimate of the population ratio R is $\hat{R}$.

Estimates of the population total, mean, ratio ($R = X/Y$) and proportion are given (from Chapter 10) by:

Mean

$$\hat{\bar{Y}} = \bar{y} = \frac{1}{n}\Sigma y_i \tag{10.1}$$

Total

$$\hat{Y} = N\bar{y} = \frac{N}{n}\Sigma y_i = \frac{1}{f}\Sigma y_i \tag{10.2}$$

Ratio

$$\hat{R} = \frac{\bar{y}}{\bar{x}} = \frac{\Sigma y_i}{\Sigma x_i} \tag{10.3}$$

Proportion

$$\hat{P} = \frac{A}{N} = \frac{a}{n} \tag{10.4}$$

The variances of the estimates are given (from Chapter 10) by:

Variance of a mean

$$v(\hat{\bar{Y}}) = \frac{1}{n(n-1)}(\Sigma y_i^2 - n\bar{y}^2) \tag{10.5}$$

Variance of a total

$$v(\hat{Y}) = N^2 v(\hat{\bar{Y}}) = \frac{N^2}{n(n-1)} (\Sigma y_i^2 - n\bar{y}^2) \qquad (10.6)$$

Variance of a ratio

$$v(\hat{R}) = \frac{1}{\bar{x}^2 n(n-1)} (\Sigma y_i^2 + \hat{R}^2 \Sigma x_i^2 - 2\hat{R}\Sigma y_i x_i) \qquad (10.7)$$

(approximately, provided that n is large)

Variance of proportion

$$v(\hat{P}) = \frac{p(1-p)}{n-1} \qquad (10.8)$$

Standard errors can be calculated by taking the square root of each variance.

Single stage design – calculations from the whole sample

The information in Table A1.1 has been collected from a sample of ten households taken from a village with a population of 200 households, by a single stage simple random sample. Calculations of sums,

Table A1.1. *Household area and production (SRS)*

Household	Area fertilized (ha)	Area of maize (ha)	Production of maize (kg)
1	0.24	0.30	426
2	0.10	0.15	174
3	0.00	0.10	0
4	0.05	0.00	0
5	0.31	0.12	209
6	0.25	0.17	375
7	0.19	0.25	313
8	0.00	0.11	104
9	0.12	0.09	126
10	0.15	0.00	0
Sum	1.41	1.29	1727
Sum of squares	0.302	0.248	520719
Mean	0.141	0.129	173
Sum of product (area × production)		343.76	

sums of squares, means and sums of product of maize area and production are given at the foot of the data.

We are required to calculate estimates and their standard errors of:

- Total fertilised area
- Mean area fertilised per household
- Mean yield rate
- Proportion of maize growers

Total fertilised area

Using equation 10.2 above

$$\hat{Y} = \frac{N}{n} \Sigma y_i$$

For the calculation we need to know the sum of observations of fertilised area, number of units in the population and number of units in the sample.

From the data above:

$$\Sigma y_i = 1.41$$

$$N = 200$$

$$n = 10$$

$$\therefore \quad \hat{Y} = \frac{200}{10} * 1.41$$

Total fertilised area = 28.2 ha

Variance of total fertilised area

Using equation 10.6 above

$$v(\hat{Y}) = \frac{N^2}{n(n-1)} (\Sigma y_i^2 - n\bar{y}^2)$$

For this calculation the only further information required is the sum of squares of fertilised area and the mean fertilised area.

$$\Sigma y_i^2 = 0.302$$

$$\bar{y} = 0.141$$

$v(\hat{Y}) = 200210 * 9\,(0.302 - 10 * 0.1412)$

Variance = 45.8

Taking the square root, the standard error of the estimate is obtained

Standard error = 6.8 ha

This can be used to calculate the confidence interval of the total or for a test of difference from another value.

Mean area fertilised per household
Using equation 10.1 above

$$\hat{Y} = \frac{1}{n}\Sigma y_i$$

From the data

$$n = 10$$

$$\Sigma y_i = 1.41$$

$$\hat{Y} = \frac{1.41}{10}$$

Mean area fertilised = 0.14 ha

Variance of mean area fertilised
Using equation 10.5 above

$$v(\hat{Y}) = \frac{1}{n(n-1)}(\Sigma y_i^2 - n\bar{y}^2)$$

From the data

$$n = 10$$

$$\Sigma y_i^2 = 0.302$$

$$\bar{y} = 0.141$$

$$v(\hat{Y}) = \frac{1}{10 * 9}(0.302 - 10 * 0.141^2)$$

Variance = 0.0012

Taking the square root, the standard error of the estimate is obtained

Standard error = 0.03 ha

Mean yield rate

Yield is a ratio variable, calculated from the sum of output divided by the sum of area. Using equation 10.3 above

$$\hat{R} = \frac{\Sigma y_i}{\Sigma x_i}$$

In this example, y refers to maize production and x to maize area. From the data above

$$\Sigma y_i = 1727 \text{ kg}$$

$$\Sigma x_i = 1.29 \text{ ha}$$

$$\hat{R} = \frac{1727}{1.29}$$

Yield rate = 1339 kg/ha

Variance of mean yield rate

Using equation 10.7 above

$$v(\hat{R}) = \frac{1}{\bar{x}^2 n(n-1)} (\Sigma y_i^2 + \hat{R}^2 \Sigma x_i^2 - 2\hat{R}\Sigma y_i x_i)$$

The additional data required now are the mean area of maize and its sum of squares, and the sum of the product of maize area and maize production, shown with the data above.

$$\bar{x} = 0.129$$

$$\Sigma y_i^2 = 520719$$

$$\hat{R} = 1339$$

$$\Sigma x_i^2 = 0.248$$

$$\Sigma y_i x_i = 343.76$$

$$v(\hat{R}) = \frac{1}{10 * 9 * 0.129^2} (520719 + 1339^2 * 0.248$$

$$- 2 * 1339 * 343.76)$$

Variance = 29895

Taking the square root, the standard error of the estimate is obtained

Standard error = 173 kg/ha

Proportion of maize growers

A proportion is calculated from the count of observations possessing a characteristic divided by the count of all observations. From the data above, eight of the ten households grow maize, households 4 and 10 do not grow maize

$$\hat{P} = p = \frac{8}{10}$$

Proportion = 0.80 or 80%

Variance of the proportion of maize growers

From equation 10.8 above

$$v(\hat{P}) = \frac{p(1-p)}{n-1}$$

From the data

$$p = 0.80$$

$$n = 10$$

$$v(\hat{P}) = \frac{0.8(1-0.8)}{9}$$

Variance = 0.02

Taking the square root, the standard error of the estimate is obtained

Standard error = 0.13

The calculations shown above are based on the whole sample, with the intention of obtaining population estimates. But very often, a characteristic is possessed by just a small proportion of the sample units, for example, cultivation of a minor or specialised crop, ownership of domestic animals, or income from specific non-farm activities. In such cases it may not be very meaningful to present population statistics, because so many observations are zero. Instead it is more useful to present statistics for the subpopulation with the characteristic. Calculations of means and ratios can be made in a similar way as for the whole sample, except that the data relate only to the subpopulation owning that characteristic.

Single stage design – calculations from a subpopulation

Using the data from Table A1.1 we now consider those households growing maize (Table A1.2).

We wish to calculate estimates and their standard errors of

- Mean area of maize
- Mean yield rate

for those households who grew maize.

Mean area of maize

Using equation 10.1 above

$$\hat{Y} = \frac{1}{n_d} \Sigma y_i$$

Table A1.2. *Maize area and production (SRS)*

Household	Area of maize (ha)	Production of maize (kg)
1	0.30	426
2	0.15	174
3	0.10	0
5	0.12	209
6	0.17	375
7	0.25	313
8	0.11	104
9	0.09	126
Sum	1.29	1727
Sum of squares	0.248	520719
Mean	0.16	216
Sum of product	343.76	

Households 4 and 10 are missing from the list because they did not grow maize

where n_d is the sample size within the subpopulation or domain of study.
From the data

$$n_d = 8$$

$$\Sigma y_i = 1.29$$

$$\hat{Y} = \frac{1.29}{8}$$

Mean area = 0.16 ha

Variance of mean area of maize
Using equation 10.5 above

$$v(\hat{Y}) = \frac{1}{n(n-1)}(\Sigma y_i^2 - n\bar{y}^2)$$

substitute n_d for n.

From the data

$$n = 8$$

$$\Sigma y_i^2 = 0.248$$

$$\bar{y} = 0.16$$

$$v(\hat{Y}) = \frac{1}{8*7}(0.248 - 8*0.16^2)$$

Variance = 0.0008

Taking the square root, the standard error of the estimate is obtained

Standard error = 0.028 ha

Mean yield rate
Using equation 10.3 above

$$\hat{R} = \frac{\Sigma y_i}{\Sigma x_i}$$

The calculation is identical to the example above for the whole sample, giving a yield rate of 1339 kg/ha

$$\Sigma y_i = 1727\ kg$$

$$\Sigma x_i = 1.29\ \text{ha}$$

$$\hat{R} = \frac{1727}{1.29}$$

Yield rate = 1339 kg/ha

Variance of mean yield rate

Using equation 10.7 above

$$v(\hat{R}) = \frac{1}{\bar{x}^2 n(n-1)} (\Sigma y_i^2 + \hat{R}^2 \Sigma x_i^2 - 2\hat{R} \Sigma y_i x_i)$$

The difference from the first example concerns the number of observations. n_d is substituted for n.

$$\bar{x} = 0.16$$

$$\Sigma y_i = 520719$$

$$\hat{R} = 1339$$

$$\Sigma x_i^2 = 0.248$$

$$\Sigma y_i x_i = 343.76$$

$$v(\hat{R}) = \frac{1}{0.16^2 * 8 * 7} (520719 + 1339^2 * 0.248 - 2 * 1339 * 343.76)$$

Variance = 31232

Taking the square root, the standard error of the estimate is obtained

Standard error = 177 kg/ha

Calculation of totals, means and ratios from a two-stage sample

The second set of examples are based on two-stage designs. Multi-staging was described in Chapter 3, and two common examples were given. The first, termed PPS/SRS makes a selection with probability proportional to size at the first stage, then a simple random sample of a constant size. This design is ideal for a situation where resident, static enumerators are used. The second design is termed SRS/SRS. Both the first and second stages are selected by simple random sample, but in order to ensure a self-weighting design, selection at the second stage is at a constant sampling fraction. It is suitable for single-visit surveys with mobile enumerators, and in situations where there is no information available about the size of first stage units which could be used for a PPS sample.

Estimation follows logically from the examples of a single-stage sample but now must take into account the multiple staging.

Using the following notation.

Observation for *j*th subunit within *i*th unit	y_{ij}
Number of second-stage subunits (households) within *i*th first-stage unit (village) in sample	m_i
Number of second-stage subunits (households) within *i*th first-stage unit (village) in population	M_i
Number of first-stage units (villages) in sample	n
Number of first-stage units (villages) in population	N
Total number of subunits (households) in survey population	H
Population total of the variable *y*	Y
Mean per second-stage subunits (overall household mean)	$\bar{\bar{Y}}$
Mean per subunit in *i*th unit in sample (household sample mean in *i*th village)	$\bar{Y}_i$
Mean per subunit in *i*th unit in population (household population mean in *i*th village)	$\bar{y}_i$
Overall mean	$\bar{y}$
Sample fraction at first stage	f_1
Sample fraction at second stage	f_2

Note: **Estimates are indicated by a caret above the symbol**

The subscript i refers to units at the first stage of selection and subscript j refers to subunits at the second stage of selection. Thus an observation would be referred to as the jth subunit in the ith unit. In the case of village first-stage units and household second-stage subunits a unit of study would be the jth household in the ith village. The village/household example is used to simplify the notation and examples.

Calculations of total, mean and ratio from two-stage PPS/SRS design

Estimates for population total, mean and ratio are set out here without derivation. Readers wishing to explore these statistics in greater depth should consult the texts in the references particularly W. G. Cochran (1977).

Total

$$\hat{\bar{Y}} = \frac{H}{mn} \sum_i \sum_j y_{ij} \tag{A1.1}$$

(In this design, $m_i = m$ is the same for all sample villages)

Mean

$$\hat{\bar{Y}} = \frac{\hat{Y}}{H}$$

$$= \frac{1}{mn} \sum_i \sum_j y_{ij} \tag{A1.2}$$

Ratio

$$\hat{R} = \frac{\hat{Y}}{\hat{X}}$$

$$= \frac{\sum_i \sum_j y_{ij}}{\sum_i \sum_j x_{ij}} \tag{A1.3}$$

The estimators for the corresponding variances are given below.

Variance of a total

$$v(\hat{Y}) = \frac{H^2}{n(n-1)m^2} \left(\sum_i \left(\sum_j y_{ij} \right)^2 - \frac{1}{n} \left(\sum_i \sum_j y_{ij} \right)^2 \right) \tag{A1.4}$$

Variance of a mean

$$v(\hat{\bar{Y}}) = \frac{1}{n(n-1)m^2} \left(\sum_i \left(\sum_j y_{ij} \right)^2 - \frac{1}{n} \left(\sum_i \sum_j y_{ij} \right)^2 \right) \tag{A1.5}$$

Variance of a ratio

$$v(\hat{R}) = \frac{H^2}{n(n-1)m^2 \hat{X}^2} \left(\sum_i \left(\sum_j y_{ij} \right)^2 + \hat{R}^2 \sum_i \left(\sum_j x_{ij} \right)^2 - 2\hat{R} \sum_i \left(\sum_j y_{ij} \right) \left(\sum_j x_{ij} \right) \right) \tag{A1.6}$$

Standard errors are given by the square root of the variance.

A maize production survey is undertaken on a sample of 25 households from a population of 9600, reported in Table A1.3. The sample design is two-stage. At the first stage five villages were selected with probability proportional to number of households. For the second stage a fixed sample of five households was drawn in each village by linear systematic sample. The data are given below. Sums are calculated for each village. Note that a sum of squares is not calculated for each village, but a square of the sum of observations is calculated.

We wish to calculate estimates and their standard errors of

- Total area of maize
- Mean area of maize
- Mean yield rate

Total area of maize

Using equation A1.1 above

$$\hat{Y} = \frac{H}{mn} \sum_i \sum_j y_{ij}$$

from the data

$$H = 9600$$
$$m = 5$$
$$n = 5$$
$$\sum_i \sum_j y_{ij} = 7.94$$

$$\hat{Y} = \frac{9600}{5 * 5} * 7.94$$

Total area = 3049 ha

Variance of total area of maize

Using equation A1.4 above

$$v(\hat{Y}) = \frac{H^2}{n(n-1)m^2} \left(\sum_i \left(\sum_j y_{ij} \right)^2 - \frac{1}{n} \left(\sum_i \sum_j y_{ij} \right)^2 \right)$$

from the data

$$\sum_i \left(\sum_j y_{ij} \right)^2 = 13.14$$

$$v(\hat{Y}) = \frac{9600^2}{5(4)5^2}\left(13.14 - \frac{1}{5} * 7.94^2\right)$$

Variance = 97926

Taking the square root, the standard error of the estimate is obtained

Standard error = 313 ha

Mean area of maize
Using equation A1.2 above

$$\hat{\hat{Y}} = \frac{1}{mn}\sum_i\sum_j y_{ij}$$

from the data

$$m = 5$$

$$n = 5$$

$$\sum_i\sum_j y_{ij} = 7.94$$

$$\hat{\hat{Y}} = \frac{1}{5 * 5}7.94$$

Mean area = 0.32 ha

Variance of mean area of maize
Using equation A1.5 above

$$v(\hat{\hat{Y}}) = \frac{1}{n(n-1)m^2}\left(\sum_i\left(\sum_j y_{ij}\right)^2 - \frac{1}{n}\left(\sum_i\sum_j y_{ij}\right)^2\right)$$

from the data

$$m = 5$$

$$n = 5$$

$$\sum_i\left(\sum_j y_{ij}\right)^2 = 13.14$$

$$\sum_i\sum_j y_{ij} = 7.94$$

Table A1.3. *Maize area and production (PPS/SRS)*

Village	Total households	Household number	Area of maize (ha)	Production of maize (kg)
A	200	1	0.21	311
		2	0.15	260
		3	0.60	807
		4	0.11	200
		5	0.00	0
Sum			1.07	1578
Square of sum			1.14	2490084
Mean			0.21	316
Product of sums			1688	
B	300	1	0.82	1010
		2	0.24	374
		3	0.09	124
		4	0.19	143
		5	0.55	329
Sum			1.89	1980
Square of sum			3.57	3920400
Mean			0.38	396
Product of sums			3742	
C	500	1	0.10	65
		2	0.24	0
		3	0.32	416
		4	0.21	196
		5	1.12	1742
Sum			1.99	2419
Square of sum			3.96	5851561
Mean			0.40	484
Product of sums			4814	
D	250	1	0.37	326
		2	0.55	747
		3	0.00	0
		4	0.24	357
		5	0.29	221
Sum			1.45	1651
Square of sum			2.10	2725801
Mean			0.29	330
Product of sums			2394	

Table A1.3. *(contd.)*

Village	Total households	Household number	Area of maize (ha)	Production of maize (kg)
E	350	1	0.66	946
		2	0.24	210
		3	0.17	263
		4	0.42	449
		5	0.05	84
Sum			1.54	1952
Square of sum			2.37	3810304
Mean			0.31	390
Product of sums			3006	
Total of five villages				
Total sample households 1600				
Sum			7.94	9580
Sum of squares of sums			13.14	18798150
Mean			0.32	383
Sum of product of sums			15644	

$$v(\hat{\hat{Y}}) = \frac{1}{5*4*5^2}\left(13.14 - \frac{1}{5}*(7.94)^2\right)$$

Variance = 0.001

Taking the square root, the standard error of the estimate is obtained

Standard error = 0.03 ha

Mean yield rate of maize

Using equation A1.3 above

$$\hat{R} = \frac{\sum_i \sum_j y_{ij}}{\sum_i \sum_j x_{ij}}$$

from the data

$$\sum_i \sum_j y_{ij} = 9580$$

$$\sum_i \sum_j x_{ij} = 7.94$$

$$\hat{R} = \frac{9580}{7.94}$$

Yield rate = 1207 kg/ha

Variance of mean yield rate
Using equation A1.6

$$v(\hat{R}) = \frac{H^2}{n(n-1)m^2\hat{X}^2}\left(\sum_i\left(\sum_j y_{ij}\right)^2 + \hat{R}^2\sum_i\left(\sum_j x_{ij}\right)^2 - 2\hat{R}\sum_i\left(\sum_j y_{ij}\right)\left(\sum_j x_{ij}\right)\right)$$

from the data

$n = 5$

$m = 5$

$H = 9600$

$\hat{X} = 3049$ (area of maize, by calculation above)

$\hat{R} = 1207$ (by calculation)

The sums of the squares of sums are:

$$\sum_i\left(\sum_j y_{ij}\right)^2 = 18798150$$

$$\sum_i\left(\sum_j x_{ij}\right)^2 = 13.14$$

and the sum of products of sums is:

$$\sum_i\left(\sum_j y_{ij}\right)\left(\sum_j x_{ij}\right) = 15644$$

Hence

$$v(\hat{R}) = \frac{9600^2}{5 * 4 * 5^2 * 3049^2} \times (18798150 + 1207^2 * 13.14 - 2 * 1207 * 15644)$$

Variance = 3500

Taking the square root, the standard error of the estimate is obtained.

Standard error = 59 kg/ha

Calculations of total, mean and ratio from two-stage SRS/SRS design

Using the same notation.

Estimates for population total, mean and ratio, with their variances, are reproduced without derivation.

Total

$$\hat{Y} = \frac{1}{f_1 f_2} \sum_i \sum_j y_{ij} \tag{A1.7}$$

Mean

$$\hat{\bar{Y}} = \frac{\sum_i \sum_j y_{ij}}{\sum_i m_i} \tag{A1.8}$$

Ratio

$$\hat{R} = \frac{\sum_i \sum_j y_{ij}}{\sum_i \sum_j x_{ij}} \tag{A1.9}$$

Variance of a total

$$v(\hat{Y}) = \frac{N}{f_1 f_2^2 (n-1)} \left(\sum_i \left(\sum_j y_{ij} \right)^2 - \frac{1}{n} \left(\sum_i \sum_j y_{ij} \right)^2 \right) + \frac{1}{f_1 f_2} \sum_i \frac{M_i}{m_i - 1} \left(\sum_j y_{ij}^2 - \frac{1}{m_i} \left(\sum_j y_{ij} \right)^2 \right) \tag{A1.10}$$

Variance of a mean

$$v(\hat{\bar{Y}}) = \frac{f_1 N}{(\Sigma m_i)^2 (n-1)} \left(\sum_i \left(\sum_j y_{ij} \right)^2 - \frac{1}{n} \left(\sum_i \sum_j y_{ij} \right)^2 \right) + \frac{f_1 f_2}{(\Sigma m_i)^2} \sum_i \frac{M_i}{m_i - 1} \left(\sum_j y_{ij}^2 - \frac{1}{m_i} \left(\sum_j y_{ij} \right)^2 \right) \tag{A1.11}$$

Variance of a ratio

$$v(\hat{R}) = \frac{N}{f_1 f_2^2 (n-1)\hat{X}^2}\left(\sum_i\left(\sum_j y_{ij}\right)^2 + \hat{R}^2\sum_i\left(\sum_j x_{ij}\right)^2 - 2\hat{R}\sum_i\left(\left(\sum_j y_{ij}\right)\left(\sum_j x_{ij}\right)\right)\right) + \frac{1}{f_1 f_2 \hat{X}^2}\sum_i \frac{M_i}{m_i - 1} \times \left(\sum_j y_{ij}^2 + \hat{R}^2\sum_j x_{ij}^2 - 2\hat{R}\sum_j y_{ij}x_{ij}\right) \quad \text{(A1.12)}$$

The information in Table A1.4 has been collected from a small survey with a two-stage design. Villages were chosen by simple random sample, because no estimate of size was available. Five villages were selected

Table A1.4. *Household information (SRS/SRS)*

Village	Household	Number of Chitumbuka speakers	Number of people in household	Total bags of rice produced	Bags marketed
A	1	5	5	10	6
	2	6	6	2	0
	3	2	2	15	15
	4	0	7	0	0
B	1	0	3	5	4
	2	0	4	10	8
	3	9	9	6	6
	4	7	8	0	0
	5	2	2	2	0
	6	11	11	12	8
C	1	1	1	4	0
	2	4	5	10	10
	3	10	10	0	0
D	1	6	7	3	0
	2	4	4	7	6
	3	0	5	10	10
	4	2	2	11	10
	5	9	9	0	0
	6	11	13	4	4
	7	4	4	6	5
	8	7	7	9	4
E	1	4	4	17	15
	2	7	7	3	0
	3	1	1	12	12
	4	11	13	11	10
	5	6	6	4	2

from a population of 30. Within each village households were selected at a constant sampling fraction of 10%. We do not know the total number of households in the population, only the number in the chosen villages.

We wish to make a number of estimates. First is the total number of Chitumbuka speakers in the survey area. The sums and sums of squares needed are given in Table A1.5.

Total number of Chitumbuka speakers

Using equation A1.7 above

$$\hat{Y} = \frac{1}{f_1 f_2} \sum_i \sum_j y_{ij}$$

from the data

$$f_1 = \frac{5}{30} = 0.17$$

$$f_2 = 10\% = 0.10$$

$$\sum_i \sum_j y_{ij} = 129$$

$$\hat{Y} = \frac{1}{0.17 * 0.10} * 129$$

Total = 7588 people

Variance of the total

Using equation A1.10 above

$$v(\hat{Y}) = \frac{N}{f_1 f_2^2 (n-1)} \left(\sum_i \left(\sum_j y_{ij} \right)^2 - \frac{1}{n} \left(\sum_i \sum_j y_{ij} \right)^2 \right)$$

Table A1.5. *Chitumbuka speakers*

Village	Sum	Square of sum	Sum of squares	Sample size
A	13	169	65	4
B	29	841	255	6
C	15	225	117	3
D	43	1849	323	8
E	29	841	223	5
Totals	129	3925	983	26

$$+\frac{1}{f_1 f_2}\sum_i \frac{M_i}{m_i - 1}\left(\sum_j y_{ij}^2 - \frac{1}{m_i}\left(\sum_j y_{ij}\right)^2\right)$$

from the data

$$N = 30$$

$$n = 5$$

$$M_i = m_i * 10 \text{ since } f_2 = 0.10$$

$$\sum_i \left(\sum_j y_{ij}\right)^2 = \text{total of squares of sums for villages} = 3925$$

$$\therefore \quad v(\hat{Y}) = \frac{30}{(0.17)(0.10)^2(5-1)}\left(3925 - \frac{1}{5} * 129^2\right)$$

$$+\frac{1}{(0.17)(0.10)}\left(\frac{40}{(4-1)}\left(65 - \frac{13^2}{4}\right)\right.$$

$$+\frac{60}{(6-1)}\left(255 - \frac{29^2}{6}\right)$$

$$+\frac{30}{(3-1)}\left(117 - \frac{15^2}{3}\right) + \frac{80}{(8-1)}\left(323 - \frac{43^2}{8}\right)$$

$$\left.+\frac{50}{(5-1)}\left(223 - \frac{29^2}{5}\right)\right)$$

Variance = 2870961

Taking the square root, the standard error of the estimate is obtained

Standard error = 1694

Next we wish to estimate the mean household size. The sums and sums of squares needed are given in Table A1.6.

Mean household size

Using equation A1.8 above

$$\hat{\bar{Y}} = \frac{\sum_i \sum_j y_{ij}}{\sum_i m_i}$$

from the data

$$\sum_i \sum_j y_{ij} = 155$$

$$\Sigma m_i = 26$$

$$\hat{\bar{Y}} = \frac{155}{26}$$

Mean household size = 6.0 people

Variance of household size

Using equation A1.11 above

$$v(\hat{\bar{Y}}) = \frac{f_1 N}{(\Sigma m_i)^2 (n-1)} \left(\sum_i \left(\sum_j y_{ij} \right)^2 - \frac{1}{n} \left(\sum_i \sum_j y_{ij} \right)^2 \right)$$

$$+ \frac{f_1 f_2}{(\Sigma m_i)^2} \sum_i \frac{M_i}{m_i - 1} \left(\sum_j y_{ij}^2 - \frac{1}{m_i} \left(\sum_j y_{ij} \right)^2 \right)$$

from the data

$$N = 30$$

$$n = 5$$

$M_i = m_i * 10$ since sampling fraction f_2 is 0.10

$\sum_i \left(\sum_j y_{ij} \right)^2$ = total of squares of sums for villages = 5587

For first village

Table A1.6. *People per household*

Village	Sum	Square of sum	Sum of squares	Sample size
A	20	400	114	4
B	37	1369	295	6
C	16	256	126	3
D	51	2601	409	8
E	31	961	271	5
Totals	155	5587	1215	26

$$\frac{M_i}{m_i - 1}\left(\sum_j y_{ij}^2 - \frac{1}{m_i}\left(\sum_j y_{ij}\right)^2\right) = \frac{40}{(4-1)}\left(114 - \frac{20^2}{4}\right)$$

and similarly for the other villages.

$$\begin{aligned}\therefore \quad v(\hat{\hat{Y}}) = {} & \frac{(0.17)30}{26^2(5-1)}\left(5587 - \frac{155^2}{5}\right) \\ & + \frac{(0.17)(0.10)}{26^2}\left[\frac{40}{(4-1)}\left(114 - \frac{20^2}{4}\right)\right. \\ & + \frac{60}{(6-1)}\left(295 - \frac{37^2}{6}\right) + \frac{30}{(3-1)}\left(126 - \frac{16^2}{3}\right) \\ & + \frac{80}{(8-1)}\left(409 - \frac{51^2}{8}\right) \\ & \left. + \frac{50}{(5-1)}\left(271 - \frac{31^2}{5}\right)\right]\end{aligned}$$

Variance = 1.56

Taking the square root, the estimated standard error is obtained

Standard error = 1.25 people

The last task is to calculate the proportion of rice produced which was marketed. The proportion marketed is a ratio, of bags marketed to bags produced.

The sums, sums of squares and sum of products are given in Table A1.7.

Table A1.7. *Calculation of sums of squares and products*

	Bags produced					Bags marketed		
Village	Sum	Square of sum	Sum of squares	Product of sums	Sum of products	Sum	Square of sum	Sum of squares
A	27	729	329	567	285	21	441	261
B	35	1225	309	910	232	26	676	180
C	14	196	116	140	100	10	100	100
D	50	2500	412	1950	334	39	1521	293
E	47	2209	579	1833	517	39	1521	473
Totals	173	6859	1745	5400	1468	135	4259	1307

Proportion marketed

Using equation A1.9 above

$$\hat{R} = \frac{\sum_i \sum_j y_{ij}}{\sum_i \sum_j x_{ij}}$$

from the data

$$\sum_i \sum_j y_{ij} = 135$$

$$\sum_i \sum_j x_{ij} = 173$$

$$\hat{R} = \frac{135}{173}$$

Proportion marketed = 0.78 or 78%

Variance of the estimate using equation A1.12 above

$$v(\hat{R}) = \frac{N}{f_1 f_2^2 (n-1)\hat{X}^2}\left(\sum_i \left(\sum_j y_{ij}\right)^2 + \hat{R}^2 \sum_i \left(\sum_j x_{ij}\right)^2 - 2\hat{R} \sum_i \left(\sum_j y_{ij}\right)\left(\sum_j x_{ij}\right)\right)$$

$$+ \frac{1}{f_1 f_2 \hat{X}^2} \sum_i \frac{M_i}{m_i - 1}\left(\sum_j y_{ij}^2 + \hat{R}^2 \sum_j x_{ij}^2 - 2\hat{R} \sum_j y_{ij} x_{ij}\right)$$

from the data

$$\hat{X} = \text{estimated total bags produced}$$

$$= \frac{173}{(0.17)(0.10)} = 10176$$

$\sum_i \left(\sum_j y_{ij}\right)^2$ = total of squares of sums for villages for bags marketed = 4259

$\sum_i \left(\sum_j x_{ij}\right)^2$ = total of squares of sums for villages for bags produced = 6859

$\sum_i \left(\sum_j y_{ij}\right)\left(\sum_j x_{ij}\right)$ = total of products of sums = 5400

For the first village:

$$\sum_j y_{ij}^2 = \text{sum of squares for bags marketed} = 261$$

$$\sum_j x_{ij}^2 = \text{sum of squares for bags produced} = 329$$

$$\sum_j y_{ij}x_{ij} = \text{sum of products} = 285$$

and similarly for the other villages.

Hence,

$$v(\hat{R}) = \frac{30}{(0.17)(0.10)^2(5-1)10176^2}(4259 + (0.78)^2 6859 - 2(0.78)5400)$$
$$+ \frac{1}{(0.17)(0.10)10176^2}\left[\frac{40}{(4-1)}(261 + (0.78)^2 329 - 2(0.78)285)\right.$$
$$+ \frac{60}{(6-1)}(180 + (0.78)^2 309 - 2(0.78)232)$$
$$+ \frac{30}{(3-1)}(100 + (0.78)^2 116 - 2(0.78)100)$$
$$+ \frac{80}{(8-1)}(293 + (0.78)^2 412 - 2(0.78)334)$$
$$\left. + \frac{50}{(5-1)}(473 + (0.78)^2 579 - 2(0.78)517)\right]$$

Variance = 0.0009

Taking the square root, the estimated standard error is obtained.

Standard error = 0.03

Two-stage designs: Calculations for a subpopulation

When considering single-stage designs, we discussed the situation where we require estimates which relate to a particular subpopulation, or domain of study, which is of interest for some reason, rather than to the whole population. A similar situation can arise when using a two-stage design. For both types of design, the calculation of the estimates themselves is very straightforwardly obtained by proceeding as if the subpopulation were the whole population, as is shown in Table A1.2. Caution is needed, however, in estimating the variances or standard errors. For some kinds of estimates, particularly those of totals, the variance calculations are not entirely straightforward, and it would be advisable to obtain specialist advice if possible, or to refer to a text on sampling theory.

Appendix 2

Detection of fabricated data

Useful techniques

Of all the types of fabrication discussed in Chapter 9 it is the crudest – the repetition of a single value – which is in some ways the most difficult to demonstrate as statistically very improbable. This is because it is expected that a yield distribution, for example, will have some degree of central tendency, and that repeated values will therefore be more likely to occur naturally in the middle of the distribution than in the tails. At the same time, it is impossible to estimate the true, central tendency of a distribution containing fabricated values.

A possible approach is to compare the suspect distribution with a distribution similar to those believed to prevail in unfabricated data. The typical form of a yield distribution, excluding cases of complete crop failure, is somewhat negatively skewed but reasonably close to a normal distribution. Typical parameters for unthreshed yield are given in Table A2.1.

These parameters have been set on the low side, as to both mean and standard deviation, in order to give tests on the conservative, or charitable, side. Higher means could, and higher standard deviations certainly

Table A2.1. *Example yield parameters*

Yield category	Low	Medium	High
Mean (kg/ha)	250	1000	10000
Standard deviation	100	400	4000
Typical crop	Cowpea	Sorghum	Yam

would, be associated with greater dispersion of the component values, decreasing the probability of genuine repetition of any given value.

The approach taken was to simulate, for each sample size of each yield category, 1000 random samples drawn from the appropriate distribution, and from these to estimate the probabilities of various numbers of repetitions of a single value. Using these probabilities it is possible to identify the number of repetitions at and above which the probability falls below a previously identified critical level. These numbers are given, for each of the three yield categories, in Table A2.2 for a critical probability level at 5% and in Table A2.3 for a critical probability level of 1%.

The use of these tables is straightforward. First, the critical probability level is selected, and hence the appropriate table. The most appropriate yield level for the crop under consideration is identified, and hence the appropriate column of the table. Then the row for the number of (non-zero) cases is found, and hence the value for the number of repetitions which exceeds the critical probability level. This or any higher number of repetitions of the same value, occurring anywhere in the yield distribution being assessed, constitutes evidence for the rejection of the data.

A similar approach can be used to estimate the probabilities of various numbers of different values, and this was also done using the simulated data. Using these probabilities, it is possible to identify the number of different values in a data set at and below which the probability falls below a previously identified critical level. These numbers are given, for

Table A2.2. *Number of repetitions of a single value of unthreshed yield for which probability is less than 5%, by simulation*

	Yield category		
No. of cases	Low	Medium	High
10	4	3	2
20	5	4	3
30	6	4	3
40	7	5	3
50	8	5	3
60	8	5	3

Note: The specified number of repetitions, or any larger number of repetitions, of any individual yield value, has a probability less than 5%.

each of the three yield categories, in Table A2.4 for a critical probability level of 5%, and in Table A2.5 for a critical probability level of 1%.

A different technique is appropriate for those cases of suspected fabrication involving a common denominator or selection for (or against) particular last digits in the subplot weight. The same technique can also

Table A2.3. *Number of repetitions of a single value of unthreshed yield for which probability is less than 1%, by simulation*

	Yield category		
No. of cases	Low	Medium	High
10	4	3	3
20	6	4	3
30	7	5	3
40	8	5	3
50	8	6	3
60	9	6	4

Note: The specified number of repetitions, or any larger number of repetitions, of any individual yield value, has a probability of less than 1%.

Table A2.4. *Total number of different values of unthreshed yield for which probability is less than 5%, by simulation*

	Yield category		
No. of cases	Low	Medium	High
10	6	8	9
20	11	15	18
30	15	23	27
40	18	30	37
50	21	36	46
60	23	42	55

Note: Any data set containing the indicated number of different values, or fewer, has a probability less than 5%.

be used when the massive repetition of a single particular value is so great as to distort the distribution of last digits in subplot weights, but the test described above specifically for repetition of a single value will usually be more sensitive in such cases.

A subplot cropcut weight, taken with a precision of 0.1 kg, is typically a two- to four-digit number, of which the last digit has very little influence on the shape of the yield distribution. Therefore, although the probability of any particular yield value being repeated depends on where in the distribution it occurs, the probability of it having any particular last digit is effectively uniform, being (0.1 × number of non-zero cases). In fact, the probability of the last digit of a subplot cropcut weight, or the penultimate digit of the resulting unthreshed yield estimate, having any particular value follows the binomial distribution. Tables A2.6 and A2.7 show the 95% and 99% confidence limits for repetitions of any given penultimate digit calculated on this basis.

The interpretation of these tables is again straightforward. Having decided on the appropriate confidence level for rejection of a data set, reference is made to the appropriate table. To assess the probability of a given degree of repetition occurring naturally, reference is then made to the nearest sample size to that of the data set under consideration, excluding zero yields; alternatively, intermediate values can be interpolated between those given in the tables. If the proportion of cases with the same penultimate digit falls outside the limits given in the appropriate

Table A2.5. *Total number of different values of unthreshed yield for which probability is less than 1%, by simulation*

	Yield category		
No. of cases	Low	Medium	High
10	5	7	8
20	10	14	17
30	14	22	27
40	17	28	36
50	20	35	44
60	22	40	54

Note: Any data set containing the indicated number of different values, or fewer, has a probability of less than 1%.

row of the left-hand column of the table, or the number of cases falls outside those in the right-hand column, the data set is rejected on grounds of probably being fabricated.

General considerations for data rejection

All the tests described above are probabilistic in nature; they can indicate when a particular data set is very highly unlikely to have been created by accurate recording of natural processes, but they cannot prove that it is absolutely impossible for the data to have arisen naturally. There is thus always the possibility of rejecting data which are in fact genuine,

Table A2.6. *95% confidence intervals for repetition of values of penultimate digits of unthreshed yields*

No. of cases	C.I. for proportion of sample with same digit	C.I. for number of cases with same digit
10	0.00–0.44	0.0–4.4
20	0.01–0.32	0.2–6.3
30	0.02–0.26	0.6–8.0
40	0.03–0.24	1.1–9.5
50	0.03–0.22	1.7–10.9
60	0.04–0.20	2.3–12.3

Extracted from Fisher & Yates *Statistical Tables*, Table VIIIi.

Table A2.7. *99% confidence intervals for repetition of values of penultimate digits of unthreshed yields*

No. of cases	C.I. for proportion of sample with same digit	C.I. for number of cases with same digit
10	0.00–0.54	0.0–5.4
20	0.00–0.39	0.1–7.7
30	0.01–0.32	0.3–9.6
40	0.02–0.28	0.7–11.3
50	0.02–0.24	1.6–12.9
60	0.03–0.24	1.6–14.4

Extracted from Fisher & Yates *Statistical Tables*, Table VIIIi.

and the first stage in any data assessment exercise using these tests is therefore to decide on the acceptable level of risk of such rejection occurring.

The criteria against which suspect data sets are to be tested are shown in Tables A2.2 to A2.5 in terms of critical probability levels, that is, they represent events with a probability less than a stated level. For example, referring to Table A2.2 for a sample size of 30 the probability of a natural occurrence of three or more repetitions of a single yield value for a high-yielding crop such as yam is less than 5%, i.e. less than one in 20. In Tables A2.6 and A2.7 the test criteria are shown in terms of confidence limits, which are simply the converse of the critical levels used in Tables A2.2 to A2.5. For example, Table A2.6 shows that, on average, 95% of samples of 30 cases will have not less than 2% and not more than 26% of cases with the same penultimate digit. A value outside this range therefore again has a probability of less than one in 20, or less than a one in 40 chance of falling outside in any given direction.

So far as assessment strategy is concerned, the important implication is that the level of critical probability of confidence interval selected is also a measure of the size of the small but positive number of genuine data sets which will be excluded because their aberrant characteristics put them outside the selected threshold of acceptance. If it is decided to exclude data sets which have characteristics indicating a less than 5% probability of their genuine origin, the probability of rejecting a genuine data set is 5% or one in 20. If a 1% critical probability level (99% confidence level) is used, 'good' data sets will have a probability of 1%, or one in 100, of being rejected.

The risk of rejecting genuine data must, however, be balanced against that of accepting fabricated data and thereby invalidating any analysis subsequently attempted. The effects of fabricated data on analytical validity are so severe that the trade-off point between the two types of risk must be set fairly low – say, at the 95% confidence level – despite the occasional injustice to field enumeration and supervision personnel which must inevitably result. In practice, the decision to reject is not often very finely balanced, fabricated data usually failing to obtain acceptance by wide margins on several criteria.

Examples of application

The use of the tests described above can be demonstrated with the data sets given in Chapter 9, Table 9.2 as examples of fabrication.

List 1

Let us suppose that, on the evidence outlined in Chapter 9, the hypothesis is formed that the data in List 1 are partly or wholly fabricated. The fabrication seems primarily to take the form of massive repetition of a single value, so our first technique, based on assessment of the probability of such repetitions, is appropriate. Let us also suppose that a critical probability level of 5% has been selected as the limit for acceptance of repetitions as being of natural origin. Reference is therefore made to Table A2.2.

The mean yield of the crop in List 1 approximates to the medium yield category, so the medium yield column in Table A2.2 is selected. This shows that for a sample of 40 non-zero cases (i.e. well over the actual sample size) the occurrence of five repetitions of the same value is an event having a probability of under 5%. The distribution in List 1, with 34 cases and five repetitions of a single value (1120 kg/ha) thus has less than a one in 20 chance of being of natural origin, and the data are accordingly rejected.

It is also possible to apply the second technique described, considering the overall number of different values in the data set. Taking, again, a critical level of 5%, reference is made to Table A2.4. This shows that for a sample of 30, 23 or fewer different values, and for a sample of 40, 30 or fewer different values, have a probability of less than 5%. The data of List 1, with 21 different values in 34 cases, are therefore rejected.

The same conclusion can be reached for the List 1 data set using the third technique described, assessment of the probability of natural repetition of the final digit of the subplot weight or the penultimate digit of the unthreshed yield. Following the procedure described above, and assuming that 95% confidence level is required for rejection of the data, reference is made to Table A2.6. The data set in List 1 has 34 non-zero values, and values from the table intermediate between those given for 30 and 40 cases are therefore appropriate. From the left-hand column of the table it is seen that the upper confidence limit for the proportion of a sample of 34 displaying any single value of the penultimate digit is between 0.26 and 0.24 – say, about 0.25; that is, if the data set has more than 25% of its values with the same penultimate digit, it should be rejected. From the right-hand column, at a sample size of 34 a number of repetitions between 8.0 and 9.5 – say, 8.7 – will equally be grounds for rejection. Inspecting the data set, it is apparent that nine out of 34, or 26%, of the values have 2 as the penultimate digit, and 13 out of 34, or 38%, have 3 as the penultimate digit. There is less than a one in forty chance of either of these events occurring naturally, and the data set is therefore assessed as having been subject to fabrication.

List 2

Let us suppose that suspicion of fabrication has fallen on the data set shown in List 2, and that again a 95% confidence level (5% critical probability level) is considered sufficient for rejection of the data set. Although the main feature of the List 2 data set is the enormous number of repeated multiples of 100, there is also a certain degree of repetition of particular values, especially the value 1200 kg/ha, which is repeated four times. It is therefore worth seeing how the data perform under the first test, for repetition of individual values.

The yield data in List 2 conform approximately to the type of the medium yield category, so the medium yield column is selected in Table A2.2. The number of non-zero cases in List 2 is 27, so the row of Table A2.2 for a sample size of 30 is a conservative (charitable) approximation. This shows that four repetitions of a single yield value is an event with a probability under 5%; the data are therefore rejected.

Applying the second test, for overall numbers of different values, we refer again to the medium yield column of Table A2.4. For a sample size of 30, 23 or fewer different values, and for a sample size of 20, 15 different values, have a probability of less than 5%. By interpolation, we have a critical level of 21 different values for a sample size of 27. The data of List 2, with 17 different values, are therefore rejected.

Applying our third test, for repeated values of the penultimate digit, for a confidence level of 95% reference is made to Table A2.6, again at a sample size of 30. At this sample size the upper confidence limit for proportion of the sample having the same value of the penultimate digit is 0.26; if more than 26% of cases show the same penultimate digit the data are rejected. Inspection of the data set shows that 23 out of 27 values, or over 85%, have 0 as their penultimate digit. This is well above the upper limit, so the data set is rejected.

List 3

The data set in List 3 shows very little repetition of individual yield values (indeed, in the original assessment the natural appearance of the frequency distribution was such that it passed an initial screening, only being detected when attention was turned to the repetition of many penultimate digits). Our first test, based on repetitions of a single value, is therefore unlikely to be conclusive. Table A2.2 (assuming retention of the 5% critical probability level) gives a rejection level of four repetitions at a sample size of 30 non-zero cases, which approximates to the actual size of 29 non-zero cases. The greatest number of repetitions of an individual yield in the List 3 data set is in fact two, so that the data are not rejected by this test. The second test is likewise not sensitive to anomalies of the kind observed in this data set.

Turning to the third test, for repeated values of the penultimate digit, reference is made to Table A2.6, where for a sample size of 30 the upper confidence limit for number of repeated values is shown as 8.0. Inspection of the data set shows that the highest number of repeated values of the penultimate digit is nine (for the value 2), which is outside the confidence limit. The data set would therefore be rejected on this criterion.

It must be remembered that it is not only a very high number of repetitions of a value which is unlikely; a very low or zero number of repetitions is also unlikely to occur naturally. Reference to Table A2.6 for a sample size of 30 shows that the lower confidence limit is 0.6 repetitions of any value of the penultimate digit. This is interpreted as meaning that any occurrence of zero repetitions of any value of the penultimate digit will be sufficient grounds for rejection of the data. In fact the data set in question has zero occurrences of five different values of the penultimate digit, indicating a very low probability indeed (about one in one hundred million) for the natural occurrence of the data set.

Glossary

Accuracy The closeness of a sample estimate to the true population value.

Bias The difference between the mean of the sample distribution of an estimator, and the true population value.

Calendar of events A list of historical events of local, national or international importance, which are likely to be known to the respondent, to help date occurrences needed for the survey, for example, accession of paramount chiefs, independence from colonial rule, years of drought.

Case study An enquiry in which a small number of study units are investigated in great detail. Usually, but not necessarily, the study units are selected purposively.

Census An investigation which covers individually every unit in the population being studied.

Closed question A question which has a number of pre-specified response categories.

Cluster sampling A type of sampling in which the primary sampling unit is usually a geographic area and consists of a group or cluster of secondary smaller units.

Coefficient of variation (CV) *Either*: The standard deviation of a set of data expressed as a proportion or percentage of its mean; *Or*: The standard error of an estimator expressed as a proportion or percentage of the estimate.

Confidence level The minimum acceptable level for the proportion of samples for which the estimate is within the specified level of precision.

Control groups Units of study which are excluded from experimental treatment, or stimulus. Under experimental conditions units would be assigned at random to treatment or control groups. For agro-economic surveys of change and development the phrase may be used to describe units not included in the development activity.

Crop output The total harvested quantity of a crop from a stated unit – field, farm, village etc.

Crop yield The total harvested quantity divided by total area, or output per unit of area.

CSV Comma separated value. A computer data file format in which consecutive data items within a data record are separated by commas.

DIF Data interchange format. A computer data file format defined by Software Arts Inc., and used by many programs to export data to or import data from other programs.

Estimate An estimated value of a population parameter, calculated from a sample.

Estimator A formula for calculating a population estimate in terms of sample observations.

Exit poll A type of enquiry in which respondents are enumerated as they leave or pass through some clearly-defined location. Usually, a form of linear systematic sampling is used to select the respondents.

Field A contiguous area of land owned or rented by a farmer.

Finite population correction (FPC) A factor by which estimates of sampling variance are multiplied, to allow for the effect of sampling from a finite rather than an infinite population. It can usually be ignored if the sampling fraction is 5% or less.

Formal sampling An expression sometimes used to distinguish the formally-structured, random sample, using a sampling frame, from the more informal sampling sometimes found in exploratory surveys, where no frame is used and the sample is not randomised.

Generalisable Relating to results which can be generalised from the sample to the population.

Hardware A general term used to describe the physical components of a computer system.

Histogram A graphical presentation of frequencies for continuous data by the areas of rectangular blocks, arranged in vertical or horizontal columns.

Holding The total area farmed by a household including fallow land. A holding may include land owned or rented by the household, plus enterprises undertaken by the household which are not land-based.

Informal sampling A type of sampling, sometimes found in exploratory surveys, where no sampling frame is used and the sample is not randomised.

Inter-penetrating subsamples Subsamples created when an existing sample is divided at random into several portions, and the fieldwork and data repeated to produce estimates of non-sampling error.

Interval data Data which are ordered, and for which the distance between adjacent items is determined, but the zero point and unit of measurement are arbitrary, for example temperature.

Intra-cluster correlation The extent to which elements in different clusters are correlated with one another. Positive intra-cluster correlation reduces the efficiency of cluster sampling, compared to simple random sampling.

Linear systematic sampling (LSS) A type of sampling in which units are selected from a frame by first taking a random starting point, and then selecting units at regular intervals thereafter.

Mean The average value of a number of observations, calculated by dividing their sum by the total number of observations.

Mean square error (MSE) The mean of the squares of the deviations of the estimates obtained from all possible samples from the true population value. This measures the combined effect of sampling error and bias.

Median The value of the middle item in an ordered set of data. Fifty per cent of observations lie above and 50% below the median.

Mode The data value which occurs with the highest frequency.

Multi-stage sample A type of sample in which, first, a sample of larger units is selected, and then from each of these a subsample of the smaller units of which they are composed is chosen. Further stages of subsampling may then follow.

Nominal data A simple classification without any expression of relative magnitude, for example, codes for male or female.

Non-sampling error That part of the deviation of a sample survey estimate from the true population value which is due to factors involved with the data collection itself, as opposed to the sample design, for example, ambiguous phrasing of questions and enumerator error in recording respondents' answers.

Open question A question which does not have a range of pre-specified categories of responses.

Ordinal data Ranked data, where the distance between adjacent items in the ranking is not determined.

Plot A subdivision of a field containing a single crop or homogeneous mixture of crops. In most cases a plot is defined to meet the needs of the surveyor. The boundary definition may not be recognised by, or meaningful to, the farmer.

Population The entirety of a defined set of units of study: the total human population of a village or a state; the total number of plants in a field; the total number of children enrolled at primary schools in a province.

Precision The closeness of a sample estimate to the mean of the sampling distribution. This relates only to sampling error, and takes no account of non-sampling error.

Probability proportional to size sampling (PPS) A type of sampling in which a unit's chance of selection is directly proportional to its size, usually measured in terms of the number of subunits of which it is composed.

Proportion The number of units possessing some characteristic, divided by the total number of units with or without that characteristic.

Proportional sampling A type of sampling in which a sample is divided into subsamples, each of a particular subdivision of the population, in proportion to the numbers of sampling units in the subdivisions.

Proxy respondent A respondent who answers questions on behalf of the intended respondent.

Purposive sampling A type of sampling in which the selected units are chosen not by chance but deliberately, in order to include units with particular characteristics, or for reasons of convenience.

Random numbers Numbers selected by chance from a specified range, in such a way that every number in that range has an equal chance of selection. Random numbers are used in random sampling, and can be obtained from tables, or by using a random number generator.

Random sample A sample selected by chance, in such a way that every unit has a known probability of being selected.

Randomise To make something random, or cause it to follow random principles.

Ratio One number or variable divided by another.

Ratio data Similar to interval data in that the data are ordered and the distance between adjacent items is determined, but there is a true zero point, for example, height or weight.

Recall period The length of time into the past from which a respondent is asked to recall information.

Reference period The period for which a respondent is asked to give information about events occurring within it.

Replication Repeating the same measurement for a number of different units, either in the same population or in the same subpopulation.

Respondent A person from whom information is obtained during data collection.

Root mean square error (RMSE) The square root of the Mean Square Error.

Sample The selected part of the population being studied.

Sample survey An investigation in which a sample of the population is studied.

Sampling distribution The distribution an estimator would have, if it were calculated for every possible sample.

Sampling error That part of the difference between a sample estimate and the true population value which arises because of estimating from a sample rather than from the whole population. It represents the uncertainty due to sampling.

Sampling fraction The proportion of the population which is included in the sample, calculated as the sample size divided by the number of units in the population.

Sampling frame A list or ordered arrangement of all the units in the population, which is used for selecting a sample.

Sampling interval The interval between the selected units in the frame, when obtaining a Linear Systematic Sample.

Sampling unit The type of unit or element which is used in selecting a sample, and which is then the basic unit to be measured, observed or interviewed.

Sampling with replacement A type of sampling in which, when a unit has been selected for the sample, it is replaced into the pool of units for selection and may be chosen again.

Sampling without replacement A type of sampling in which, when a unit has been selected for the sample, it is removed from the pool of units for selection and cannot be chosen again.

Self-weighting sample A type of sample in which all units have an equal chance of being selected, so the sample values do not need to be weighted to compensate for different probabilities of selection.

Simple random sampling A type of sampling in which every unit is assigned a number, and the sample selected by using random numbers.

Single-stage sampling A sample which is selected in one stage only, without further subsampling.

Software A general term used to describe the set of instructions given to a computer to perform operations – the computer program.

Standard deviation (SD) A measure of the dispersion of a parameter's distribution. It is calculated by taking the square of the deviation of each value from the mean, summing these squares, dividing the sum by the number of elements, and then taking the square root.

Standard error (SE) The standard deviation of the sampling distribution of an estimate. This is a measure of the sampling variability or precision of the estimate.

Stratification The division of the population into subpopulations, which are then sampled independently. Usually, the subpopulations are defined so as to be more homogeneous than the whole population, in order to reduce the overall sampling variability.

Two-stage sample A multi-stage sample with two stages; a primary and secondary sampling unit.

Unit of study An entity or event defined to be the subject for the collection of data. Commonly a household or a person, or an event such as a market transaction.

Universe The whole population being investigated, not just that which is being sampled.

Variance The square of the standard deviation of a distribution.

Weight The factor by which a particular sample value has to be multiplied in order to compensate for different probabilities of selection.

References

APMEPU (1982). Methodology Manual. Kaduna, Nigeria: Agricultural Projects Monitoring, Evaluation and Planning Unit. (mimeo)

Barnett, T. (1977). *The Gezira Scheme.* London: Frank Cass.

Bauchi State ADP (1984). *Crop Production and the Food Balance in Bauchi State.* Bauchi, Nigeria: BSADP Evaluation Unit.

Booker, W. (1980). Household survey experience in Africa. *LSMS Working Paper No. 6,* World Bank.

Buzan, T. (1974). *Use Your Head.* British Broadcasting Corporation Publications.

Carruthers, I. and Chambers, R. (1981). Rapid appraisal for rural development. *Agricultural Administration,* 8, 401–422.

Casley, D. J. and Lury, D. A. (1981). *Data Collection in Developing Countries.* Oxford: Clarendon Press.

Casley, D. J. and Kumar, K. (1987). *Project Monitoring and Evaluation in Agriculture.* Baltimore, MD: Johns Hopkins.

Casley, D. J. and Kumar, K. (1988). *The Collection, Analysis and Use of Monitoring and Evaluation Data.* Baltimore, MD: Johns Hopkins.

Caswell, F. (1989). *Success in Statistics,* 2nd edn. London: John Murray.

Chambers, R. (1983). *Rural Development, Putting the Last First.* Harlow, Essex: Longmans.

Chander, R., Grootaert, C. and Pyatt, G. (1980). Living standards surveys in developing countries. *LSMS Working Paper No. 1,* World Bank.

Chapman, M. and Mahon, B. (1986). *Plain Figures.* HMSO.

Cleave, J. H. (1974). *African Farmers: Labor Use in the Development of Smallholder Agriculture.* New York: Praeger.

Cochran, W. G. (1977). *Sampling Techniques,* 3rd edn. New York: James Wiley.

Coleman, G. (1982). Labour data collection in African traditional agricultural systems: a methodological review and examination of some Nigerian data. *Development Studies Occasional Paper No. 18.* University of East Anglia.

Collinson, M. P. (1972). *Farm Management in Peasant Agriculture.* New York: Praeger.

Cooper, B. M. (1975). *Writing Technical Reports.* Harmondsworth, England: Pelican.

FAO (1978). Collecting statistics on agricultural population and employment. *Economic and Social Development Paper No. 7.* Rome: FAO.

FAO (1982). The estimation of crop areas and yields in agricultural statistics. *Economic and Social Development Paper No. 22.* Rome: FAO.

Farrington, J. (1975). Farm surveys in Malawi. *Development Study No. 16,* University of Reading, Department of Agricultural Economics and Management.

Fisher, R. A. and Yates, F. (1963). *Statistical Tables for Biological, Agricultural and Medical Research,* 6th edn. Edinburgh: Oliver & Boyd.

Glewwe, P. (1990). Improving data on poverty in the Third World. *PRE Working Paper.* World Bank.

Greeley, M. (1987). *Post Harvest Losses, Technology and Employment: The Case of Rice in Bangladesh.* Boulder, CO: Westview Press.

Grootaert, C. and Marchant, T. (1991). The social dimensions of adjustment priority survey. *SDA Working Paper No. 12.* World Bank.

Hildebrand, P. E. (1981). Combining disciplines in rapid appraisal: the Sondeo approach. *Agricultural Administration,* 8, 407–422.

ILCA (1981). *Low-level Aerial Survey Techniques.* Addis Ababa: International Livestock Centre for Africa.

Jahnke, H. E. (1982). *Livestock Production Systems and Livestock Development in Tropical Africa.* Keil, Germany: Kieler Wissenschaftsverlag Vauk.

Lipton, M. and Moore, M. (1972). The methodology of village studies in less developed countries. *IDS Discussion Paper No. 10,* University of Sussex.

McCracken, J. A. (1988). A working framework for rapid rural appraisal: lessons from a Fiji experience. *Agricultural Administration and Extension,* 29, 163–183.

Miles, M. B. and Huberman, A. M. (1984). *Qualitative Data Analysis.* Beverley Hills, CA: Sage.

Molokwu, C. and Poate, C. D. (1981). Yield Estimation in Yam and Sorghum: A Study of Crop Cutting from Subplots. Kaduna, Nigeria: Agricultural Projects Monitoring, Evaluation and Planning Unit. (mimeo).

Murphy, J. and Marchant, T. J. (1988). *Monitoring and Evaluation in Extension Agencies.* World Bank.

Norman, D. W. (1970). An economic study of three villages in Zaria province: Part 1. *Samaru Miscellaneous Paper No. 33.* Samaru, Nigeria: Institute for Agricultural Research.

Panse, V. G. (1958). Some comments on the objectives and methods of the 1960 World Census of Agriculture. *Bulletin of the International Statistical Institute,* 36, 222–227.

Poate, D. (1988). A review of methods for measuring crop production from smallholder producers. *Experimental Agriculture,* 24, 1–14.

Poate, C. D. and Casley, D. J. (1985). *Estimating Crop Production in Development Projects. Methods and their Limitations.* World Bank.

Putt, S. N. H., Shaw, A. P. M., Woods, A. J., Tyler, L. and James, A. D. (1988). *Veterinary Epidemiology and Economics in Africa.* Addis Ababa: International Livestock Centre for Africa.

Rees, D. G. (1985). *Essential Statistics.* London: Chapman and Hall.

Robinson, H. (1981). *Database Analysis and Design.* England: Chartwell–Bratt.

Salmen, L. F. (1987). *Listen to the People.* Oxford: Oxford University Press.

Scott, C. (1985). *Sampling for Monitoring and Evaluation.* World Bank.

Scott, C. and Amenuvegbe, B. (1990). Effect of recall duration on reporting of household expenditures. *SDA Working Paper No. 6.* World Bank.

Snedecor, G. W. and Cochran, W. G. (1967). *Statistical Methods,* 6th edn. Ames: Iowa State University Press.

Srinivas, M. N., Shah, A. M. and Ramaswamy, E. A. (eds) (1979). *The Fieldworker and the Field: Problems and Challenges in Sociological Investigations.* Delhi: Oxford University Press. (Cited in Chambers, 1983).

Sudman, S. and Bradburn, N. M. (1982). *Asking Questions.* San Francisco, CA: Jossey-Bass.

Sukhatme, P. V. (1954). *Sampling Theory of Surveys with Applications.* Ames: Iowa State University Press.

Swift, J. (1981). Rapid appraisal and cost-effective participatory research in dry pastoral areas of West Africa. *Agricultural Administration,* 8, 485–492.

The Economist (1986). *Pocket Style Book.* London: The Economist Publications.

Tukey, J. W. (1977). *Exploratory Data Analysis.* Addison–Wesley.

Verma, V., Marchant, T. and Scott, C. (1988). Evaluation of Crop-cut Methods and Farmer Reports for Estimating Crop Production. London: Longacre Consultants. (mimeo)

World Bank (1989). *Structural Adjustment and Poverty: A Conceptual, Empirical and Policy Framework.* Washington, DC: World Bank.

Yates, F. (1949). *Sampling Methods for Censuses and Surveys.* London: Griffin.

Zarkovitch, S. S. (1966). *Quality of Statistical Data.* Rome: FAO.

Index

Page numbers in italic indicate a table or a diagram.